Urban Water Cycle Processes and Interactions

Urban Water Series – UNESCO-IHP

ISSN 1749-0790

Series Editors:

Čedo Maksimović
Department of Civil and Environmental Engineering
Imperial College
London, United Kingdom

Alberto Téjada-Guibert
International Hydrological Programme (IHP)
United Nations Educational, Scientific and Cultural Organization (UNESCO)
Paris, France

Urban Water Cycle Processes and Interactions

Jiri Marsalek, Blanca Jiménez-Cisneros, Mohammad Karamouz, Per-Arne Malmquist, Joel Goldenfum and Bernard Chocat

UNESCO Publishing

Taylor & Francis
Taylor & Francis Group

Cover illustration

Aalborg Municipality, Denmark, aerial view of Aalborg, with kind permission.

Published jointly by

The United Nations Educational, Scientific and Cultural Organization (UNESCO)
7, place de Fontenoy
75007 Paris, France
www.unesco.org/publishing

and

Taylor & Francis The Netherlands
P.O. Box 447
2300 AK Leiden, The Netherlands
www.taylorandfrancis.com – www.balkema.nl – www.crcpress.com
Taylor & Francis is an imprint of the Taylor & Francis Group, an informa business, London, United Kingdom.

Typeset by Charon Tec Ltd (A Macmillan company), Chennai, India
Printed and bound in Hungary by Uniprint International (a member of the Giethoorn Media-group), Székesfehévár.

ISBN UNESCO, paperback: 978-92-3-104060-3
ISBN Taylor & Francis, hardback: 978-0-415-45346-2
ISBN Taylor & Francis, paperback: 978-0-415-45347-9
ISBN Taylor & Francis e-book: 978-0-203-93246-9

Urban Water Series: ISSN 1749-0790

Volume 2

The designations employed and the presentation of material throughout this publication do not imply the expression of any opinion whatsoever on the part of UNESCO or Taylor & Francis concerning the legal status of any country, territory, city or area or of its authorities, or the delimitation of its frontiers or boundaries.
The authors are responsible for the choice and the presentation of the facts contained in this book and for the opinions expressed therein, which are not necessarily those of UNESCO nor those of Taylor & Francis and do not commit the Organization.

British Library Cataloguing in Publication Data
A catalogue record for this book is available from the British Library

Library of Congress Cataloging-in-Publication Data
Urban water cycle processes and interactions / J. Marsalek ... [et al.].
     p. cm. — (Urban water series — UNESCO ihp ; v. 2)
   Includes bibliographical references and index.
   ISBN 978-0-415-45346-2 (hardcover : alk. paper) — ISBN 978-0-415-45347-9 (pbk. : alk. paper) —
ISBN 978-0-203-93246-9 (e-book)
   1. Municipal water supply. 1. Marsalek, J. (Jiri), 1940-

TD209.U73 2007
628.109173'2—dc22
                                                                            2007032489

Foreword

Continuing urbanization leads to ever increasing concentrations of population in urban areas. General statistics indicate that currently just over half of the world's population lives in urban areas, and in some countries this proportion reaches 90% or more. The process of urbanization is particularly marked in developing countries, which account for a disproportionately high number of megacities with many millions of inhabitants with sprawling periurban areas. Consequently, the issue of urban environmental sustainability is becoming increasingly critical, because urbanization and its associated environmental impacts are occurring at an unprecedented rate and scope.

These concerns have long been recognized by UNESCO's International Hydrological Programme (IHP), which has addressed the role of water in urban areas, effects of urbanization on the hydrological cycle and water quality, and many aspects of integrated water management in urban areas, and is developing integrated approaches to deal with today's acute problems in this domain.

This book presents results of one of these studies; its main focus is on the assessment of anthropogenic impacts on the urban hydrological cycle and the urban environment, including processes and interactions in the urban water cycle. The need for this study follows from the fact that effective management of urban waters should be based on a scientific understanding of anthropogenic impacts on the urban hydrological cycle and the environment. Such impacts vary broadly in time and space, and need to be quantified with respect to the local climate, urban development, cultural, environmental and religious practices, and other socio-economic factors.

This publication, which is part of a series of urban water management books produced within the framework of the Sixth Phase of IHP (2002–2007) is the main output of the project on processes and interactions in the urban water cycle by IHP. In order to address the broad range of conditions in urban water management, UNESCO established a working group for this study with representatives of various professional backgrounds and experience from various climatic regions, whose deliberations and joint efforts resulted in this book. The role of Mr Jiri Marsalek as the driving force in the composition of the book and as lead editor is amply recognized. The book has been produced under the responsibility of Mr J. Alberto Tejada-Guibert, Deputy Secretary of IHP. We are grateful to all the contributors and the editors for their hard work, and are confident that the conclusions, recommendations and case studies contained in this volume will prove to be of value to urban water management practitioners, policy-and decision makers and educators throughout the world.

András Szöllösi-Nagy
Secretary of UNESCO's International Hydrological Programme (IHP)
Director of UNESCO's Division of Water Sciences
Deputy Assistant Director-General for the Natural Sciences Sector of UNESCO

Contents

List of Figures

List of Tables

Acronyms

ABR	Anaerobic baffled reactor
ADB	Asian Development Bank
ASR	Aquifer storage recovery
BMP	Best management practice
BOD	Biochemical oxygen demand
CEC	Commission of European Communities
COD	Chemical oxygen demand
CSO	Combined sewer overflow
DAF	Dissolved air flotation
DDE	Dichlorodiphenylethylene
DDT	Dichlorodiphenyltrichloroethane
DNAPL	Dense non-aqueous phase liquid
DO	Dissolved oxygen
DOC	Dissolved organic carbon
EC	European Community
EC	*Escherichia coli*
ECOSAN	Ecological sanitation
EDS	Endocrine disrupting substance
EEA	European Environment Agency
EEC	European Economic Community
EPA	Environmental Protection Agency (USA)
ESSA	East, South and South-East Asia
ET	Evapotranspiration
FCU	Faecal coliform units
GTZ	Deutsche Gesellschaft für Technische Zusammenarbeit GmbH (German society for technical cooperation)
HO	Helminth ova
IHP	International Hydrological Programme
INSA	Institut National des Sciences Appliquées de Lyon
IOL	Indicator organism limit
IPH	Instituto de Pesquisas Hidráulicas
IWMED	Institute for Wetland Management and Ecological Design (India)
kPa	Kilo-Pascal (unit of pressure)
MENA	Middle East and North Africa
MOE	Ontario Ministry of Environment (Canada)

MPN	Most probable number
MSF	Multi-stage flash evaporation
MST	Microbial source tracking
N	Nitrogen
NBT	Nash bargaining theory
NO_x	Oxides of nitrogen
NRC	National Research Council (US)
NSW	New South Wales (Australia)
NURP	Nationwide Urban Runoff Program
P	Phosphorus
PAH	Polycyclic aromatic hydrocarbon
PFU	Plaque forming unit
PPCP	Pharmaceuticals and personal care products
RO	Reverse osmosis
RTC	Real-time control
SAT	Soil aquifer treatment
SBR	Sequential batch reactor
SCS	Soil Conservation Service (US)
SO_x	Oxides of sulphur
SSA	Sub-Saharan Africa
TDS	Total dissolved solids
THM	Trihalomethanes
TN	Total nitrogen
TP	Total phosphorus
TSS	Total suspended solids
TWCP	Tehran Wastewater Collection Project
UASB	Upflow anaerobic sludge blanket
UFRGS	Universidade Federal do Rio Grande do Sul (Brazil)
UK	United Kingdom of Great Britain and Northern Ireland
UNEP	United Nations Environmental Programme
UNESCO	United Nations Educational, Scientific and Cultural Organization
UN-HABITAT	United Nations Centre for Human Settlements
UNICEF	United Nations Children's Emergency Fund
US	United States
USLE	Universal soil loss equation
UV	Ultraviolet
UWC	Urban water cycle
UWWE	Urban wastewater effluents
VOC	Volatile organic compound
VSS	Volatile suspended solids
WC	Water closet
WHO	World Health Organization
WPCF	Water Pollution Control Federation (US)
WWTP	Wastewater treatment plant

Glossary

adiabatic cooling a natural atmospheric process whereby an air mass cools due to lower pressures as it rises, while maintaining the same volume. This effect can cause water vapour to condense and form rain or snow in the presence of condensation nuclei.

aggregate water use water use for all purposes aggregated by a territorial (geographical) unit, e.g., county, or by a hydrologic unit, e.g., a catchment, or groundwater aquifer; *see also* disaggregate water use.

bedload sediment in almost continuous contact with the bed of a watercourse, carried by rolling, sliding or hopping.

benthic species/communities aquatic organisms which spend all or part of their life cycle at the sediment-water interface.

berm low earthen wall constructed with sloping sides and arranged to contain, direct or control surface runoff in stormwater management.

biocenosis a set of organisms, plants and animals living in ecological balance which is based on the chemical, physical and biological conditions in the local environment.

bioengineered soils soils amended by mixing natural soils with organic materials (to improve the soil structure and water storage capacity) and stabilised and protected against erosion by vegetation.

biofiltration combined physical and chemical processes of filtration and adsorption with the uptake and processing of nutrients by attached micro-organisms.

bioretention area a vegetated surface depression designed to collect, store and infiltrate runoff; where needed, the underlying soil layer is replaced with a bioengineered soil.

biosphere parts of the Earth where life exists; they comprise the lower part of the atmosphere, the hydrosphere and the Earth's crust in contact with these.

biotype a naturally existing population comprising individuals with the same genetic makeup.

blackwater sanitary wastewater produced from a combination of faeces and urine.

capillary tension the measure of the forces of attraction between a water molecule and the clay surface, representing the sum of the adhesive and cohesive forces or the capillary forces.

chelation a chemical process of forming or joining together metallic cations with certain organic compounds.

convective storms   storms resulting from daytime heat build-up over land masses, characterised by short bursts of intense rainfall and high winds. The intense rainfall often results in considerable runoff and danger of flooding.

disaggregate water use   water use for individual purposes, including public supply, irrigation, industrial, thermoelectric power, mining, livestock and aquaculture categories for a particular water supply utility unit; see also aggregate water use.

distributed systems   wastewater treatment systems which are non-centralised. These systems are usually small, inexpensive, uncomplicated and are designed to treat locally wastewater generated by a small number of households.

ecohydrology   a sub-discipline of hydrology that focuses on ecological processes occurring within the hydrological cycle and strives to utilise such processes for enhancing environmental sustainability.

elution   *syn* leaching; separation of adsorbed compounds by continuous washing with a liquid (eluent).

evapotranspiration   process by which water is transferred to the atmosphere from the soil by evaporation and from the vegetation by transpiration.

exfiltration   loss of water in a drainage system to surrounding soils and groundwater as a result of percolation or adsorption processes.

ferrule-based charges   a method of charging for costs of water on a fixed basis (i.e., charging fees per a ferrule of certain size) rather than on the basis of water actually used.

fire flow   the amount of water that should be available in a water distribution network for municipal fire protection.

greywater   domestic wastewater generated from dish washing, laundry and bathing. It does not include any blackwater materials.

heat island effect   increase in air temperature in urban areas resulting from the warming effect of hard surfaces, e.g., asphalt and concrete.

hydraulic conductivity   property of a porous medium which, according to Darcy's law, relates the specific discharge to the hydraulic gradient.

hydrograph   a graphical representation of stage, i.e., water depths above some datum, or discharge as a function of time.

hydrologic components   (of the hydrological cycle)   major components of the hydrological cycle including water sources (imported water or precipitation), hydrological abstractions (interception, depression storage, evaporation and evapotranspiration, infiltration), water storage (soil moisture, groundwater), and water transport by interflow, groundwater flow, and surface runoff.

hydrological cycle   *syn.* water cycle; succession of stages through which water passes from the atmosphere to the earth and returns to the atmosphere: evaporation from the land or sea or inland water, condensation to form clouds, precipitation, interception, infiltration, percolation, runoff and accumulation in the soil or in bodies of water, and re-evaporation.

hydrological regime   variations in the state and characteristics of a water body which are regularly repeated in time and space and which pass through seasonal or other phases.

hydrometeorological variable   a variable describing weather or water characteristics, e.g., wind, solar radiation, air temperature, precipitation, water temperature, water depth, water velocity, and water quality.

hydrosphere   all the components of water on the Earth, including water vapour in clouds, ice caps and glaciers, and water in oceans, seas, lakes, rivers and aquifers.

impound   to store water in reservoirs (impoundments).

improved water supply   supply of drinking water, defined in terms of the technology and levels of service as more likely to provide safe water than unimproved technologies, including household connections, public standpipes, boreholes, protected dug wells, protected springs, and rainwater collection (after WHO).

integrated management   a planning and operational process in which interested parties, stakeholders and regulators reach general agreement on the best mix of conservation, sustainable resource use and economic development and diversification.

interflow   1) that portion of the precipitation which has not passed down to the water table, but is discharged from the area as subsurface flow into stream channels. 2) flow of water from ephemeral zones of saturation by moving through the upper strata of a geological formation at a rate much in excess of normal baseflow seepage.

lithosphere   the upper layer of the earth, including the Earth's crust and upper mantle.

lumped soil characteristics   soil characteristics aggregated over some unit volume.

megacity   a city with more than 10 million inhabitants.

microclimate   the climate in a localized area, which differs from that of the surrounding region.

particulates   tiny solid or liquid particles which are suspended in the air. Most particulates fall into the 10 nm to 100 μm diameter size range.

pelagic communities   aquatic organisms which spend all or part of their life-cycle swimming or floating in the water column. These types of organisms include algae and zooplankton.

phreatic level   a natural groundwater table (level).

pillar-based ecosystem approach   a strategy for integrated management of land, water and living resources promoting conservation and sustainability, which is based on consideration of three basic pillars – society, living environment and economy.

regression analysis   statistical method developed to investigate the interdependence or relationship between two or more measurable variates.

rugosity   roughness of river bed.

sanitary landfill   site for disposal of solid wastes which has been engineered to prevent discharge of leachates into the groundwater. This is typically achieved through an impermeable layer under the landfill and a leachate collection and treatment system.

sub-potable   water which has not been treated to drinking water standards.

swale   depression or ditch used to collect and divert runoff. Water and nutrients gradually infiltrate into the surrounding soils and create a fertile growing area.

thermocline   *syn.* metalimnion; layer in a thermally stratified body of water across which the temperature gradient is at a maximum.

time series analysis   evaluation and interpretation of a set of measurements, usually taken at regular intervals, determining the way in which these data vary along the time dimension.

transpiration   process by which water from vegetation is transferred into the atmosphere in the form of vapour; *see also* evapotranspiration.

vadose zone   *syn* unsaturated zone, zone of aeration; subsurface zone above the water table in which the interstices are filled with air and water, and the water pressure is less than atmospheric.

virtual water   a measure of indirect consumption of water in the production of agricultural or industrial products requiring significant quantities of clean water in their production.

water pressure district   a part of the water supply area served by a water distribution system and characterized by a certain water pressure.

water-soil interface   a common boundary between water and soils.

Acknowledgements

Many colleagues have contributed to the preparation of this report and their contributions are gratefully acknowledged. In particular, the work of the following is acknowledged:

- UNESCO Secretariat: Mr J.A. Téjada-Guibert, officer in charge of Urban Water activities of the International Hydrological Programme VI (IHP-VI) and Deputy Secretary of IHP; Mr C. Maksimović, adviser for the IHP Urban Water component; Ms B. Radojevic, consultant; and Mr W. H. Gilbrich, consultant.
- Mr W.E. Watt, Emeritus Professor, Queen's University, Kingston, Ontario, Canada, who served as an external editor of the final report.
- Mr Q. Rochfort, Ms J. Dziuba and Mr P. McColl, National Water Research Institute, Burlington, Ontario, Canada, who produced the print-ready version of the report.
- All the members of the Working Group: for the study of the urban water cycle processes and interactions for this project, and in particular, those who provided written materials.

The Working Group for the study of the urban water cycle processes and interactions comprised the following members:

- **Gamal Abdo**, Department of Civil Engineering, University of Khartoum, Khartoum, Sudan.
- **Bernard Chocat**, INSA Lyon, Lyon, France, contributed to Chapter 4.
- **Joel Goldenfum**, IPH/UFRGS, Porto Allegre, Brazil, contributed to Chapters 2 and 3.
- **K.V. Jayakumar**, Water and Environment Division, Regional Engineering College, Warangal, India.
- **Blanca Jiménez-Cisneros**, Environmental Engineering Department, Institute of Engineering, Universidad Nacional Autónoma de Mexico, Mexico City, Mexico, contributed to Chapters 3 and 4.
- **Mohammad Karamouz**, School of Civil Engineering, University of Tehran, Tehran, Iran, contributed to Chapters 2 and 3.
- **Per-Arne Malmquist**, Chalmers University of Technology, Goteborg, Sweden, contributed to Chapter 3.
- **Jiri Marsalek**, National Water Research Institute, Burlington, Ontario, Canada, contributed to Chapters 1–4 and provided report integration and editing.

Chapter 1

Urban water cycle

1.1 INTRODUCTION

An urban population demands great quantities of energy and raw materials, as well as the removal of waste, some of which turns into environmental pollution. Indeed, all key activities of modern cities – transportation, electricity supply, water supply, waste disposal, heating, supply of services, manufacturing, etc. – are characterized by these problems. Thus, concentration of people in urban areas dramatically alters material and energy fluxes in the affected areas, with concomitant changes in landscape, altered fluxes of water, sediment, chemicals and micro-organisms, and increased release of waste heat. These changes then impact on urban ecosystems, including urban waters and their aquatic ecosystems, and result in their degradation. Such circumstances make provision of water services to urban populations highly challenging, particularly in megacities, which are defined as the cities with 10 million or more inhabitants. Yet the number of these megacities keeps growing, particularly in the developing countries, and this further exacerbates both human health and environmental problems. The growth of the number of megacities is illustrated in Table 1.1, listing megacities in 1975 and 2003, and predictions for 2015.

While the adverse effects of urbanization are fairly well known and widely publicized, there are also positive aspects and advantages of living in well-managed cities, including important opportunities for economic and social development, a modern style of living with high female labour force participation and indicators of good levels of general health, well-being and literacy, and a limited ecological footprint (Cohen, 2006). However, the management of large urban entities is one of today's greatest challenges (Maksimovic and Tejada-Guibert, 2001). With respect to continuing urbanization, it is also important to note that the current demographic data and planning projections are highly uncertain, that most of the current growth of urban areas is caused by rural–urban migration and the transformation of rural settlements into towns and cities, and finally, that most of the expected growth will not occur in the largest cities (megacities), but in smaller secondary cities and towns in the developing world, where poverty rates are higher and general services are inadequate (Cohen, 2006).

Conflicting demands on resources require integrated management of the urbanization process, which is a most challenging task. Within this complex setting, this report focuses on the management of urban waters, recognizing that effective management of urban waters should be based on a scientific understanding of impacts that human activity has on the urban hydrological cycle and the environment, and the means of

Table 1.1 **Megacities with more than 10 million people**

1975	2003	2015
Tokyo, Japan (26.6)	Tokyo, Japan (35.0)	Tokyo, Japan (36.2)
New York, USA (15.9)	Mexico City, Mexico (18.7)	Mumbai, India (22.6)
Shanghai, China (11.4)	New York, USA (18.3)	Delhi, India (20.9)
Mexico City, Mexico (10.7)	Sao Paulo, Brazil (17.9)	Mexico City, Mexico (20.6)
	Mumbai, India (17.4)	Sao Paulo, Brazil (20.0)
	Delhi, India (14.1)	New York, USA (19.7)
	Calcutta, India (13.8)	Dhaka, Bangladesh (17.9)
	Buenos Aires, Argentina (13.0)	Jakarta, Indonesia (17.5)
	Shanghai, China (12.8)	Lagos, Nigeria (17.0)
	Jakarta, Indonesia (12.3)	Calcutta, India (16.8)
	Los Angeles, USA (12.0)	Karachi, Pakistan (16.2)
	Dhaka, Bangladesh (11.6)	Buenos Aires, Argentina (14.6)
	Osaka-Kobe, Japan (11.2)	Cairo, Egypt (13.1)
	Rio de Janeiro, Brazil (11.2)	Los Angeles, USA (12.9)
	Karachi, Pakistan (11.1)	Shanghai, China (12.7)
	Beijing, China (10.8)	Metro Manila, Philippines (12.6)
	Cairo, Egypt (10.8)	Rio de Janeiro, Brazil (12.4)
	Moscow, Russian Federation (10.5)	Osaka-Kobe, Japan (11.4)
	Metro Manila, Philippines (10.5)	Istanbul, Turkey (11.3)
	Lagos, Nigeria (10.1)	Beijing, China (11.1)
		Moscow, Russian Federation (10.9)
		Paris, France (10.0)

Source: after Marshall, 2005.

mitigation of such impacts, and should take full account of the socio-economic system. Urbanization impacts vary broadly in time and space, and need to be quantified with respect to the local climate, urban development, engineering and environmental practices, cultural religious practices and other socio-economic factors.

Analysis of urban water management should be based on the urban water cycle (UWC), which provides a unifying concept for addressing climatic, hydrologic, land use, engineering and ecological issues in urban areas. Furthermore, it was felt that the analysis of the UWC would be conducive to a later examination of modern approaches to water management in urban areas, including sustainable development, total urban water cycle management, low-impact development and ecohydrology. These approaches based on water conservation make use of integrated management measures, including integrated management and reuse of stormwater, groundwater and wastewater.

The report that follows represents the first step of a comprehensive project and aims to develop a schematic representation of the UWC, including its environmental components, and identify the major fluxes of water, sediment, chemicals, micro-organisms and heat, with reference to urban waters in both the developed and developing countries. Such a scheme may be presented in many variations reflecting various climatic conditions, both current and future (i.e. considering climate change). In subsequent study phases, it is expected that these fluxes will be quantified and described by water balance/quality models approximating such processes. Connections between urban development and these fluxes will be established, and principles for low-impact developments and restoration of the existing areas will be established. Some of the intermediate steps/results in the overall study include:

- identification of the components of the UWC and the effects of urbanization on water resources
- quantification of the imprint of human activities on the urban hydrological cycle and its interaction with the environment under the present and future development scenarios
- understanding of the processes at the urban water–soil interface, including the interaction of water and soil, with particular reference to soil erosion, soil pollution and land subsidence
- hydrological, ecological, biological and chemical processes in the urban water environment of sustainable cities of the future
- assessment of the impact of urban development, land use and socio-economic changes on the availability of water supplies, aquatic chemistry (anthropogenic) pollution, soil erosion and sedimentation, and natural habitat integrity and diversity
- assessment of the preventive and mitigation measures available for dealing with urban water problems.

The final product of this activity should be a guidance manual on alterations in the UWC and the environment caused by human activities, with reference to various climatic zones and potential climate changes. This manual should advance:

- the understanding of processes that take place in the urban environment, and of the interactions of natural suburban, rural and urban environments for the successful analysis, planning, development and management of urban water systems
- development of innovative analytical tools for addressing the problems of spatial and temporal variability
- assessment of the potential effects of climate variations and changes on urban water systems.

1.2 URBAN WATER CYCLE CONCEPT

One of the most fundamental concepts in hydrology, and indeed in water resource management, is the hydrologic cycle (also referred to as the water cycle), which has been speculated on since ancient times (Maidment, 1993). There is some diversity of definitions of the hydrological cycle, but generally it is defined as a conceptual model describing the storage and circulation of water between the biosphere, atmosphere, lithosphere and hydrosphere. Water can be stored in the atmosphere, oceans, lakes, rivers, streams, soils, glaciers, snowfields and groundwater aquifers. Its circulation among these storage compartments is caused by such processes as evapotranspiration, condensation, precipitation, infiltration, percolation, snowmelt and runoff, which are also referred to as the water cycle components.

The combined effects of urbanization, industrialization and population growth affect natural landscapes and the hydrological response of watersheds. Although many elements of the natural environment are affected by human activities with respect to

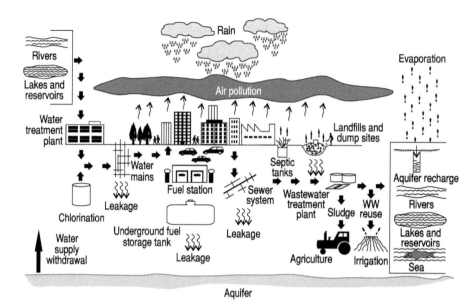

Figure 1.1 Urban water cycle

pathways and hydrologic abstractions (or sources of water), the principal structure of the hydrological cycle remains intact in urban areas. However, the cycle is greatly modified by urbanization impacts on the environment and the need to provide water services to the urban population, including water supply, drainage, wastewater collection and management, and beneficial uses of receiving waters. Thus, it was noted that the hydrological cycle becomes more complex in urban areas, because of many influences and interventions (McPherson, 1973; McPherson and Schneider, 1974); the resulting 'urban' hydrological cycle is then called the urban water cycle (UWC). The UWC is shown pictorially in some detail in Figure 1.1 and schematically in Figure 1.2, which displays just the major components and pathways.

The UWC provides a good conceptual and unifying basis for studying the water balance (also called the water budget) and conducting water inventories of urban areas, catchments, developments or sites. In such studies, the above listed major components of the hydrological cycle are assessed for certain time periods, with durations exceeding the time constants of the system to filter out short-term variability. Water balances are generally conducted on seasonal, annual or multi-year bases (van de Ven, 1988), and in planning studies, such balances are projected to future planning horizons.

The water management approach based on mass balances (water and material budgets) is particularly important for assessing the performance of urban water systems (Gumbo, 2000), for urban planning (i.e. providing water services to growing populations), and for coping with extreme weather and climatic variations and climate change. In fact, the understanding of water balances is essential for integrated management of urban water, which strives to remediate urban population pressures and impacts by intervention (management) measures, which are applied in the total management of the UWC (Lawrence et al., 1999). On a broader spatial scale of an urbanized river basin, integrated urban water

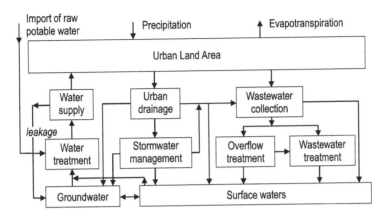

Figure 1.2 Urban water cycle – main components and pathways

management can yield such benefits as maintenance of normal streamflows, flood control, conservation and development of water resources, conservation and rehabilitation of an ecosystem, pollution control and improvement of the thermal regime (Imbe et al., 1997).

Thus, studies of the water, sediment and chemical balance help establish and quantify the UWC, by addressing such issues as: verification of pathways in the cycle; quantifying flows and fluxes of sediment and chemicals along the pathways; assessing component variations; and assessing impacts of climatic, population and physiographic changes on the UWC. Examples of urban water balances have been offered by many authors, including Hogland and Niemczynowicz (1980), Grimmond and Oke (1986), van de Ven (1988), Stephenson (who presented a summary of six studies by other authors; 1996), Herrmann and Klaus (1996) and Gumbo (2000). It should be noted that the development of urban water budgets was found useful in both highly developed countries (see for example Imbe et al., 1997) and developing countries (Gumbo, 2000). A brief description of principal components of the UWC follows.

Two main sources of water are recognized in the UWC: the municipal water supply and precipitation. Municipal water is often imported from outside the urban area or even from another catchment in widely varying quantities that reflect local water demands and their management. This water may bypass some pathways in the UWC; it is brought into the urban area and distributed within it; some fraction is lost to urban groundwater, and the rest is used by the population, converted into municipal wastewater, and eventually returned to surface waters. The second source, precipitation, generally follows a longer route through the water cycle. It falls in various forms over urban areas, is subject to hydrologic abstractions (including interception, depression storage and evapotranspiration), partly infiltrates into the ground (contributing to soil moisture and recharge of groundwater) and is partly converted into surface runoff, which may be conveyed to receiving waters by natural or artificial conveyance systems. With various success and accuracy, flow components were quantified for urban areas in studies of urban water balances (Hogland and Niemczynowicz, 1980).

Besides these clearly established (intentional) linkages among the various water conveyance and storage elements, unintentional ones may also develop (e.g. water main leaks, sewer exfiltration) and have to be addressed in water management.

In addition to flow components of the UWC, attention needs to be paid to the fluxes of materials and energy conveyed by air, water or human activities. In general, these processes are less well known and quantified than those dealing with water only, and their description in urban areas is complicated by numerous remote and local sources and high variability in time and space. With respect to atmospheric pollution – conveyed in wet form with precipitation and dry form as gases and particulates – Novotny and Olem (1994) identified the major pollutants as acidity (originating from nitrogen and sulphur oxides emitted from fossil-fuel combustion), trace metals, mercury and agricultural chemicals (particularly pesticides and herbicides). These chemicals may fall directly into receiving waters, or be deposited on catchment surfaces and subject to scouring and transport into receiving waters during wet weather.

Other pollution sources include inappropriate land use activities and poor housekeeping, including transportation, construction activities, use of building materials, road maintenance, attrition or elution or corrosion of hard surfaces, soil erosion, urban wildlife (particularly birds) and pets and deficient solid waste collection, among others. Besides being deposited directly into urban waters (which is generally of secondary importance because of the small water surface areas), these materials may be washed off and transported by urban runoff as dissolved or suspended pollutant loads, or as a bedload. During this transport, depending on hydraulic conditions, settling and re-suspension of pollutants takes place on the catchment surface and in pipes, and biological and chemical reactions also occur. These processes are often considered to be more intense in the initial phase of the storm (first flush effect); however, due to the temporal and spatial variability of rainfall and runoff flow, first flush effects are more pronounced in conveyance systems with pipes rather than on overland flow surfaces.

While past studies of urbanization and water management, particularly in developed countries, focused on science and engineering, there is growing recognition of the importance of the social conditions and links between the socio-economic system and the water and the environment (Lundqvist et al., 2001). Furthermore it is recognized that sustainable solutions to water-related problems must reflect the cultural (emotional, intellectual and moral) dimensions of people's interactions with water. Culture is a powerful aspect of water resources management. Water is known as a valuable blessing in most of the arid or semi-arid countries and most religions. There are two cultural aspects that have direct impacts on water resources management in urban areas: urban architecture and people's lifestyle.

Traditional architecture in urban areas often reflects the climate characteristics of the area. However, the traditional architecture in many large cities is being replaced by modern 'Western' architecture because of population increase and globalization, with concomitant changes in urban hydrology. The density of the population and buildings, rainwater collection systems, material used in construction, and wastewater collection systems are among the major factors that cause changes in the urban hydrologic cycle.

Lifestyles in urban areas affect the hydrologic cycle through changes in domestic water demands. Domestic water use per capita and water use in public areas such as parks and green areas are the main characteristics that define the lifestyle in large cities. Even though economic factors are important for determining these characteristics, the pattern of water use, tradition and culture have more significant effects on the lifestyle in urban areas.

1.3 TOTAL MANAGEMENT OF THE URBAN WATER CYCLE

The concept of the UWC demonstrates the connectivity and interdependence of urban water resources and human activities, and the need for integrated management. To meet this need, the concept of total UWC management was introduced in Australia, and further elaborated on by Lawrence et al. (1999). The basic water management categories encompassed in this approach include:

- Reuse of treated wastewater, as a basis for disposing of potential pollutants, or a substitute for other sources of water supply for sub-potable uses.
- Integrated stormwater, groundwater, water supply and wastewater-based management, as the basis for:
  - economic and reliable water supply
  - environmental flow management (deferment of infrastructure expansion, return of water to streams)
  - urban waterscape/landscape provision
  - substitution of sub-potable sources of water (wastewater and stormwater reuse)
  - protection of downstream waters from pollution.
- Water conservation (demand management) based approaches, including:
  - more efficient use of water (water-saving devices, irrigation practices)
  - substitute landscape forms (reduced water demand)
  - substitute industrial processes (reduced demand, water recycling).

The concept of total management of the UWC is presented here just to illustrate the importance of the UWC in integrated urban water management, recognizing that other compatible environmental management approaches also exist, including ecohydrology or a pillar-based ecosystem approach. While many of these measures have been practised in the past, what has been missing was the understanding of the linkages among the various components, and the implication of the practices for the long-term quality of groundwater, soils and environmental flows.

Finally, it should be emphasized that the concept of the UWC and of its total management applies to all climatic, physiographic, environmental and socio-cultural conditions, and the levels of development, with appropriate modifications. Indeed, the examples provided by Stephenson (1996), Imbe et al. (1997) and Gumbo (2000) cover vastly varying urban areas, including those in Australia, Canada, China, Japan, Mexico, Russia, South Africa, Sweden and Zimbabwe. Naturally, depending on local circumstances, different measures may be given different priorities, but the general principle of identifying the main sources of water, sediments, chemicals and biota, the applicable pathways or changes, and intervention measures, serving the integrated management of natural resources, remain the same and will be explored in the following chapters.

The discussion of UWC starts with a general introduction of the UWC concept in Chapter 1, followed by discussions of hydrological components of the UWC in Chapter 2, urban infrastructure and water services in Chapter 3, and urbanization effects on the environment in Chapter 4.

Chapter 2

Urban water cycle hydrologic components

Urbanization contributes to changes in the radiation flux and the amounts of precipitation, water evaporation and evapotranspiration and infiltration into soils, and consequently causes changes in the hydrological cycle. The effects of large urban areas on local microclimate have long been recognized, and occur as a result of changes in the energy regime, air pollution and air circulation patterns caused by buildings and/or transformation of land cover.

In the past, changes in the rainfall–runoff components of the hydrologic cycle, caused by urbanization, have received most attention and can be summarized as follows:

- transformation of undeveloped land into urban forms (including transportation corridors), with the concomitant changes in land cover and catchment surface imperviousness and hydrologic abstractions
- increased energy release (e.g. greenhouse gases, waste heat, heated surface runoff)
- increased demands on water supply (municipal and industrial).

Figure 2.1 shows rainfall–runoff components of the hydrologic cycle. Each of these components and the related processes are briefly explained in the following section; detailed descriptions can be found elsewhere (Viessman et al., 1989).

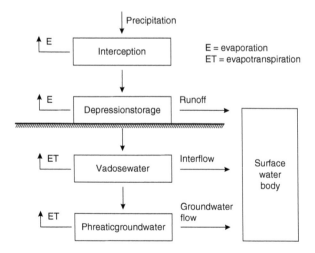

Figure 2.1 **Rainfall–runoff components of the hydrologic cycle**

Further discussion of urbanization impacts on water resources is presented in Chapters 3 and 4.

2.1 WATER SOURCES

Urban areas require large quantities of water for their sustenance, far in excess of what can be supplied by local precipitation. Consequently, additional sources of water are needed. Two main sources in urban areas, municipal water supply and precipitation, are recognized and discussed in this section.

2.1.1 Municipal water supply

Municipal water is often imported from outside the urban area (or even outside the watershed in which the urban area is located), in quantities ranging from 50 to 700 L/capita/day, as demanded by municipal water users. Municipal water use is usually categorised as residential, commercial, industrial and 'other' use, where 'other' includes water lost through leakage, unaccounted-for water uses (e.g. fire-fighting and distribution system flushing), and water not assigned to the first three categories. Thus, the quantity of imported water depends largely on the population served by municipal water supply systems, and on institutional, commercial and industrial activities. Import of water (particularly from other catchments) represents a major influence on the UWC; with typical urban water uses being non-consumptive, most of the imported water is discharged into groundwater (contributing to rising groundwater tables) or local receiving bodies as wastewater effluents (with major impacts on such water resources). Under some circumstances, some of the sub-potable water demands can be supplied by reused or recycled water, thus conserving potable water sources (see Section 3.4.10). Furthermore, the water provided through municipal water supply systems may be locally consumed, or exported as virtual water in various products, or used in ground irrigation, or converted into wastewater, which may or may not be treated prior to discharge into receiving waters. A more detailed discussion of water supply is presented in Chapter 3.

2.1.2 Precipitation

The second important source of water is precipitation, which occurs in greatly varying quantities depending on local climate. The effects of large urban areas on local microclimate have long been recognized (Geiger et al., 1987); they occur as a result of changes in the energy regime, air pollution and air circulation patterns, which are caused by buildings, land transformations and release of greenhouse gases. These factors contribute to changes in the radiation balance and the amounts of precipitation and evaporation, and consequently to changes in the hydrologic cycle. Such effects, with respect to changes in annual precipitation, air temperatures and evaporation rates are also described in Section 4.3.

2.1.2.1 Climatic aspects

Climate is defined as the long-term behaviour of the weather in a region. The hydrologic processes in different climates are affected by the hydrometeorological variables, which attain different ranges of magnitude in different climates. There are four world climate categories, corresponding to subtropical, continental, rain-shadow and cool coastal arid lands.

Subtropical areas (e.g. Sahara, Arabia, Australia and Kalahari) are characterized by clear skies with high temperatures. The summers are hot and the winters are mild, so the seasonal contrasts are evident, with low winter temperatures due to freezing. Convective rainfalls develop only when moist air invades the region.

Continental interior areas (e.g. arid areas of Asia and the western United States) have seasonal temperatures ranging from very cold in winter to very hot in summer. Snow can occur. Rainfall in the summer is unreliable in this climate.

Rain-shadow areas (mountain ranges such as the Sierra Nevada, the Great Dividing Range in Australia and the Andes in South America) are characterized by conditions similar to those in continental areas with diverse behaviour, but their climatic conditions are not as extreme as in the continental interior areas.

Cool coastal areas (e.g. the Namib Desert on the south-western coast of Africa and the Pacific coast of Mexico) have reasonably constant climatic conditions with a cool humid environment. When temperate inversions are weakened by upward-moving moist air, thunderstorms can develop.

Arid climate is one of the more important climate types of the world. Rozanov (1994) suggested that the major cause of aridity was explained through the global atmospheric circulation patterns, with certain local effects imposed by topography; Thompson (1975) listed four main processes explaining aridity: (a) high pressure, (b) wind direction, (c) topography, and (d) cold ocean currents.

The majority of semi-arid and arid regions are located between latitudes 25 and 35 degrees, where high pressure causes warm air to descend, resulting in dry, stable air masses. Aridity induced by orographic causes is common in North and South America, where high mountain ranges are perpendicular to the prevailing air mass movements. These air masses are cooled as they are forced up the mountains, reducing their water holding capacity. Most of the moisture is precipitated at high elevations of the windward slopes. The relatively dry air masses warm up as they descend on the leeward side of the mountain ranges, increasing their water-holding capacity and reducing the chance of any precipitation. This orographic aridity is referred to as the rain-shadow effect (Dick-Peddie, 1991). In central Asia, the region's positioning in the continent, where distance from oceans lessens the chance of encountering moisture-laden air masses, is the cause of the semi-arid and arid conditions. Cold ocean currents cause the coastal arid regions of Chile and Peru and the interior part of northern Argentina; cold ocean currents in close proximity to the coast supply dry air that comes on shore, but as the mass is forced up the mountain sides there is no moisture to be lost as the air mass cools. The desert climate is another important arid climate of the world.

2.1.2.2 Urban precipitation

Precipitation represents one of two primary water inputs to the UWC and is derived from atmospheric water. Recognizing that large amounts of water may accumulate in clouds without precipitation, the processes of condensation and precipitation are sometimes considered individually. Among the various causes of condensation, the most important is dynamic or adiabatic cooling, which produces nearly all precipitation.

The condensation of water vapour into droplets occurs on condensation nuclei, whose occurrence is related to air pollution. Large urban areas affect the local microclimate as a result of changes in the energy regime, air pollution (providing condensation nuclei), air circulation patterns and releases of greenhouse gases (Marsalek et al., 2001). Earlier studies have shown that total annual precipitation in, or downwind of, large industrialized cities is generally 5–10% higher than in the surrounding areas, and for individual storms this increase in precipitation can be as high as 30%, particularly on the downwind side of large metropolitan areas (Geiger et al., 1987). Further changes are expected as a result of climate change, with global circulation models predicting either increasing or decreasing precipitation, depending on the specific location, and greater climatic variability with more pronounced extremes (Van Blarcum et al., 1995).

Among the various forms of precipitation, convective storms with high rainfall intensities are particularly important for the design of urban minor drainage elements or infrastructure (the sizing of conveyance elements); cyclonic precipitation may be more important for design of major drainage and storage facilities.

While climates were traditionally considered as either non-varying or changing very slowly, recent research on greenhouse effects indicates some imminent climate changes. Solar radiation reaching the earth is partly absorbed by it, but a substantial part is reflected back into space. The heat absorbed is radiated by the earth as infrared radiation. Greenhouse gases such as water vapour, carbon dioxide, methane and nitrous oxides absorb the infrared radiation and in turn re-radiate it in the form of heat. The amount of greenhouse gases in the atmosphere is increasing due to anthropogenic activities (Houghton et al., 1996). Thus, the emission of greenhouse gases contributes to climate change. It is forecasted that increased greenhouse gas concentration will lead to increased average temperature, by 3–5°C in some regions by the year 2050. As concentrations of greenhouse gases increase even more, further climate changes can be expected. In urban areas which are affected by global climate changes, some researchers suggest that the magnitude of the effects of changing climate on water supplies may be much less important than changes in population, land cover, technologies, economics or environmental regulations (Lins and Stakhiv, 1998).

Air temperature is also of interest in studies of precipitation, because it determines the form of precipitation (e.g. rain or snow). The urban heat island effect increases air temperatures over urban areas by as much as 4–6°C, compared with surrounding localities (see Section 4.3.1). These thermal phenomena then explain higher evaporation rates (by 5–20%) in urban areas (Geiger et al., 1987) and other related effects.

2.2 HYDROLOGIC ABSTRACTIONS

The most important component of the UWC with respect to drainage and flood protection is stormwater runoff. To determine runoff, one needs to consider water input (rainfall or snowmelt), hydrologic abstractions (sometimes called losses) and the routing of net water input in the catchment. Such routing is strongly affected by storage, which modifies the inflow hydrograph. A brief overview of such processes follows.

A significant fraction of precipitation is returned to the atmosphere by evaporation or evapotranspiration, depending on local landscape and water resources. The remaining water may infiltrate into the ground (recharging groundwater), or be converted into runoff and streamflow. Generally, rainwater infiltration in urban areas is reduced by the high imperviousness of urban areas, and this contributes to increased runoff and higher risk of flooding and erosion in receiving streams (Marsalek, 2003a). Reduced hydrologic abstractions and increased surface runoff are recognized as typical impacts of urbanization on the hydrologic cycle (Leopold, 1968). Furthermore, urban runoff becomes polluted during overland flow and transport in storm or combined sewers, and consequently affects the water quality of the receiving waters (Marsalek, 2003a). Therefore, during the last thirty years, stormwater management has been introduced, with the main goal of reducing anthropogenic impacts on the hydrologic cycle and the mobilization and transport of sediments and pollutants. Typical stormwater management measures are discussed in Chapter 3.

2.2.1 Interception

Interception is defined as that part of water input that wets, and adheres to, above-ground objects (e.g. tree canopy) until it evaporates and returns to the atmosphere (Viessman et al., 1989). Water abstractions by interception are particularly important in vegetated (forested) catchments, where the amount intercepted depends on the species, age and density of the vegetation, storm event characteristics and the season of the year (Geiger et al., 1987). Interception abstractions occur early during rainstorms and quickly diminish. In urban areas with little tree cover, interception is insignificant and often neglected. Although there are formulae for calculating interception as a function of rainfall and vegetation characteristics (Chow, 1964), the estimated interception is often included in the initial abstraction and deducted from the storm rainfall (Geiger et al., 1987). Traditional urban development, with high imperviousness and little vegetation or tree cover, reduces interception and its importance in urban runoff analysis.

2.2.2 Depression storage

Depression storage (also called surface storage) accounts for water that is trapped in small depressions on the catchment surface and retained until it infiltrates or evaporates. In some hydrology handbooks, wetting abstractions (i.e. water used for the initial wetting of the catchment surface) are combined with depression storage and called the initial abstraction (Geiger et al., 1987). Depression storage depends on the characteristics of the catchment surface, including the type of surface and its slope. On impervious urban surfaces, depression storage ranges from 0.2 mm (smooth asphalt pavement) to 2.8 mm (an average value for small urban areas); on pervious surfaces, depression storage ranges from 0.5 mm (bare clay) to 15 mm (wooded areas and open fields). The relative significance of depression storage depends on the storm rainfall (or snowmelt): the larger the rainfall, the less significant is depression storage in stormwater runoff calculations. A detailed listing of depression storage values for more than twenty-five types of surfaces can be found in Geiger et al. (1987).

2.2.3 Evaporation and evapotranspiration

Evaporation is the process occurring along the water–air or soil–air interface by which water in liquid or solid state transforms into water vapour escaping into the atmosphere. Higher rates of energy consumption and higher air temperatures in cities contribute to higher rates of evaporation in urban areas (by 5–20%) (Geiger et al., 1987). Transpiration is the process of vaporization of water at the surface of plant leaves after the soil water has been transported through the plant (Overton and Meadows, 1976). For simplification, the process of transpiration is sometimes combined with evaporation from water and soil surfaces into evapotranspiration, which can be estimated by the Penman equation (Viessman et al., 1989). Furthermore, it is commonly assumed that water supply to these processes is not limited, which permits the treatment of evapotranspiration at its potential rate (Geiger et al., 1987). Land use changes in urban areas lead to a reduced extent of green areas in cities and thereby contribute to reduced total transpiration from trees and vegetation. While evaporation and evapotranspiration are important in water budget calculations (note that in the Sahelian zone, the daily evaporation can be as high as 15 mm/day), during urban stormwater runoff, both abstractions are rather small and justifiably neglected.

2.2.4 Infiltration

Infiltration is the process of water movement into the soil under gravity and capillary forces. Through this process, shallow aquifers are recharged and, by discharging to surface waters, contribute to streamflow during dry periods. Two basic approaches to describing infiltration are a soil physics approach that relates infiltration rates to detailed soil properties (e.g. hydraulic conductivity, capillary tension and moisture content), and a hydrological approach, which is parametric and utilizes lumped soil characteristics to estimate infiltration rates. The latter approach is commonly used in urban runoff calculations. For more information on infiltration calculations methods commonly used in urban runoff modelling (the Horton, Green-Ampt, Philip and Holtan approaches) see Viessman et al. (1989).

Compared with natural areas, infiltration rates decrease in urban areas because of the following factors:

- increased imperviousness of urban catchments (pavements, rooftops, parking lots, etc.)
- compaction of soils in urban areas
- presence of an artificial drainage system that provides for quick removal of ponded water, without allowing water enough time to infiltrate into the ground.

The first in-depth analysis of the urban hydrological cycle (by Leopold, 1968) noted that increased imperviousness of urban catchments contributed to lower infiltration and thereby to reduced groundwater recharge, reduced interflow/baseflow and higher rates of surface runoff. Higher rates of runoff then contribute to a higher incidence of flooding. However, recent studies indicate that urbanization may result in a net gain in overall groundwater recharge, mostly because of losses from water supply mains, leaking sewer systems and stormwater infiltration (Lerner, 2004).

2.2.5 Lumped hydrologic abstractions

In some empirical procedures for runoff calculation, hydrological abstractions are lumped into empirical coefficients. Examples of such approaches include the runoff coefficient, which applies to the runoff peak flow, the Φ-index applied to pervious areas and accounting for interception, evaporation, wetting, depression and infiltration abstractions (Chow, 1964), and the runoff curve number method of the Soil Conservation Service (SCS) (US Department of Agriculture, SCS, 1975). These lumped methods are commonly used in practice, because they avoid difficulties with determining hydrologic abstractions for various types of cover and their variation in time, but at the cost of increasing uncertainties in computed results.

2.3 WATER STORAGE

Water infiltrating into the ground contributes to soil moisture and to groundwater recharge.

2.3.1 Soil moisture

Large parts of urban areas are covered by impervious surfaces and materials (roads, parking lots, roofs) which contribute to the high imperviousness of downtown areas and, by reducing the pervious areas, decrease infiltration and evapotranspiration. Consequently, the vadose zone differs significantly from that in natural areas.

Another factor contributing to changes in soil moisture in urban areas is the partial removal of topsoil during development and construction, and changes in soil structure resulting from the use of heavy machinery. After development, less topsoil may be returned and lower soil layers may have been compacted; consequently there is reduced soil moisture storage and greater need for irrigation. Some stormwater management practices (discussed in Section 3.3.2.2) attempt to reverse this process by using bioengineered soils in urban areas to restore soil moisture storage capacity and infiltration.

2.3.2 Urban groundwater

Urbanization affects not only surface waters but also both the quantity and quality of groundwater. The state of groundwater then impacts on the water balance and resources of an urban area and on the operation of the urban infrastructure, including storm, sanitary and combined sewers, stormwater management facilities and sewage treatment plants. A detailed analysis of urban groundwater and its pollution can be found in Lerner (2004); a brief introduction of the groundwater issues is included here for the sake of completeness.

Groundwater can be characterized according to its vertical distribution into two zones, the zone of aeration and the zone of saturation (below the water table). The zone of aeration is divided, from top to bottom, into the soil–water, vadose and capillary zones (Todd, 1980). In urban areas, groundwater interacts with surface waters and urban infrastructure, and is further affected by land use activities. An Internet-based Urban Groundwater Database (www.utsc.utoronto.ca/~gwater/IAHCGUA/UGD/) indicates that urban groundwater issues and problems vary, depending on the climate, urban area, land use activities, environmental practices and other local conditions. In drier climates and developing countries, the typical trend is toward over-exploitation of groundwater for municipal and industrial water supplies; this leads to the lowering of the groundwater table, land subsidence, saltwater intrusion in coastal areas, and groundwater pollution. In developed countries, urban aquifers are generally not used for water supply; however, the water tables may still decline because the high imperviousness of urban areas may mean there is insufficient recharge. Less commonly, urban water tables may be rising due to low withdrawal of groundwater and leakage from water mains (e.g. Nottingham, UK). In all urban areas, the occurrence of groundwater pollution from various sources has been reported (Lerner, 2004).

2.4 INTERFLOW AND GROUNDWATER FLOW

Reduced infiltration due to the high imperviousness of urban catchments should contribute to smaller interflows. However, the situation is complicated in urban areas by another source of water discharged into shallow aquifers: leakage from water distribution and wastewater collection networks. Leakage from water mains is particularly important, because such pipes are pressurized. Even low water losses, expressed as 15% of volume input, provide water volumes equivalent to groundwater recharge by several hundred millimetres of rainfall annually (Lerner, 2004). Exfiltration from leaky sewer pipes depends on the relative positions of sewers and the groundwater table; this determines the direction of water transport and may vary over time. Leaky sewers in dry soils will function as a source of groundwater, but leaky sewers below the groundwater table will drain aquifers and convey this flow to sewage treatment plants. Finally, other sources of inputs to groundwater are stormwater infiltration facilities, which are applied in modern stormwater management. Such facilities include porous or permeable pavements, permeable access-holes, drainage swales, and infiltration wells, trenches and basins.

There may be also an influx of groundwater into the urban area, or into municipal sewers (Hogland and Niemczynowicz, 1980), and such waters may contribute to increased volumes of municipal sewage, which are treated at the local plant and the effluent discharged into receiving waters.

2.5 STORMWATER RUNOFF

Changes of runoff regime represent one of the most significant impacts of urbanization (see also Section 4.4.1.1). Urbanization affects surface runoff in three ways:

- by increasing runoff volumes due to reduced rainwater infiltration and evapotranspiration
- by increasing the speed of runoff, due to hydraulic improvements of conveyance channels
- by reducing the catchment response time and thereby increasing the maximum rainfall intensity causing the peak runoff discharge.

Thus urbanization changes the catchment hydrologic regime. These changes were quantified in the literature, with the mean annual flood increasing from 1.8 to 8 times, and the 100-year flood increasing from 1.8 to 3.8 times, due to urbanization (Riordan et al., 1978). Stormwater direct runoff volume increased for various return periods up to six times. In general, the magnitude of such increases depends on the frequency of storms, local climate and catchment physiographic conditions (soils, degree of imperviousness, etc.), as partly illustrated in Figure 2.2. Figure 2.3 then shows two runoff hydrographs

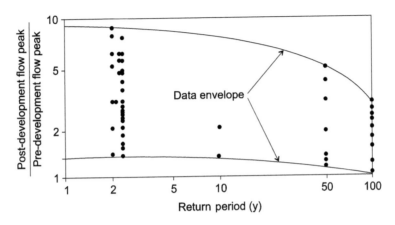

Figure 2.2 Effects of watershed development on runoff peaks

Source: Marsalek, 1980.

Note: The ratio of the post to pre-development peaks indicates the increase in runoff flow due to development; full dot symbols indicate the results reported in the literature.

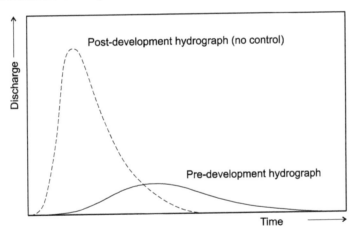

Figure 2.3 Runoff hydrograph before and after urbanization

from the same catchment, before and after urban development. The figure demonstrates changes in the runoff hydrograph caused by urbanization. As discussed later in Section 3.3.2.2, the post-development runoff peak can be controlled by storage or other measures. Note that storage reduces the peak, but not the volume of runoff, which contributes to increased runoff flows over extended time periods, with concomitant effects on channel erosion in downstream areas (see Section 4.4.1.2).

2.6 NATURAL DRAINAGE: URBAN STREAMS, RIVERS AND LAKES

The final components of the UWC are sinks, in the form of receiving waters that are elements of the natural drainage system. Two types of receiving waters are commonly recognized: surface waters and groundwater. In both cases, there are conflicts arising from multiple water uses. Receiving waters generally provide beneficial water uses, including source water for water supplies, fishing, recreation and ecological functions (e.g. aquatic habitat), but they also serve to transport, store and purify urban effluents that convey pollution. Similar conflicts have been reported for groundwater, which may serve for water supply as well as (often unintentionally) for disposal of some pollutants. Thus, to protect downstream water uses, it is necessary to manage urban effluents with respect to their quantity and quality, in order to lessen their impact on water resources.

Wastewater from various municipal sources, including residential, commercial, industrial and institutional areas, is collected by sewers or open drains and conveyed to treatment facilities, or discharged directly into receiving waters. Depending on local climate and population density, on an annual basis, the municipal effluent volume may exceed the volume of stormwater runoff from the urban area.

In urban drainage design, urban streams are considered as elements of the major drainage system, and are often modified to accommodate increased flows resulting from urbanization. The situation concerning urban lakes is similar, with urbanization changing their hydrological regime. In most cases, however, the main water management challenge is dealing with the water quality impairment (particularly siltation of streams and eutrophication of lakes), as discussed further in Chapter 4.

2.7 NEEDS FOR URBAN WATER INFRASTRUCTURE

Extensive changes of the hydrological regime in urban areas have been historically managed by building an urban infrastructure, starting with water supply aqueducts, followed by stormwater and sewage collection, and eventually sewage treatment plants. Such systems provide water services, including water supply, drainage and sewage management, and in turn interactively affect the hydrological cycle in urban areas. For example, import of drinking water into urban areas changes the urban water budget, particularly through leakage from water mains into urban aquifers. Increased stormwater runoff has to be managed in urban areas, sometimes by source controls, but more often by enhancing the conveyance capacity of natural channels and by building new ones and underground sewers.

These measures are implemented to manage local flooding in urban areas, but they also contribute to faster hydrological response in urban catchments and further increases in peak stormwater flows. Finally, most of the water imported into and used in urban areas is transformed into wastewater, which is discharged into receiving waters and further increases demands on their capacity, with respect to conveyance, storage and self-purification. In developed countries with long traditions of providing urban water services, infrastructure systems have been built over centuries and do provide good services to urban dwellers, as long as their operation and maintenance is adequately funded. As an alternative to these 'central' systems, new distributed systems are currently being promoted; these may represent an attractive alternative in developing countries without central systems, or funds to build and maintain them. The issues of urban infrastructure and provision of water services are addressed in Chapter 3.

Chapter 3

Urban water infrastructure

The urban population places high demands on water services, and at the same time it has major impacts on water resources in the affected areas. The issues of planning and delivery of water services are addressed in this chapter; the associated impacts on water resources are dealt with in Chapter 4.

When considering water services, it has to be recognized that the solutions are strongly affected by the socio-economic conditions in individual countries and, consequently, there is a need to account for two sets of conditions – those in developed and in developing (or less developed) countries. Consequently, the presentation of material in this chapter (and its sections) starts with general considerations that often reflect the situation in developed countries, and then addresses the additional challenges encountered in developing countries.

3.1 DEMANDS ON WATER SERVICES IN URBAN AREAS

Urban areas are highly dynamic and complex entities. They require various resources – including water, food, energy and raw materials – and produce wastes which need to be safely disposed of. With respect to urban water management, such a continuous movement, use and disposal of energy and materials can be visualized schematically as flows of water, food and wastes into and out of urban areas, or as ecocycles of water and nutrients supporting urban areas (Figure 3.1).

The sustainable operation of urban areas requires the provision of sustainable water services. Historically, the main components of urban water systems and the provision of related water services, including water supply, drainage, sewage collection and treatment, and receiving water uses, were addressed separately. Their interactions were often disregarded or underestimated; such an approach is obviously untenable under current conditions. Consequently, integrated approaches to water management, including sustainable development, have evolved. Sustainable development was defined by Brundtland (1987) as 'development that meets the needs of the present without compromising the ability of future generations to meet their own needs', and contextually contains two key concepts: (a) the concept of 'needs', and in particular the essential needs of the world's poor, to which overriding priority should be given, and (b) the idea of limitations imposed by the state of technology and social organization on the environment's ability to meet present and future needs. The goal of sustainable development requires implementation approaches, among which the ecosystem

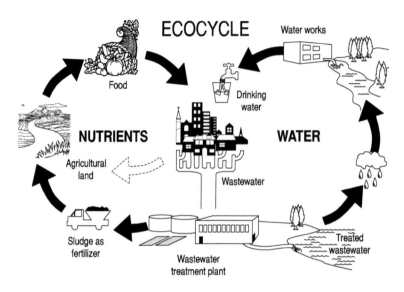

Figure 3.1 Ecocycles of water and nutrients supporting urban areas

approach is particularly common. This attributes equal importance to three basic environmental planning and management pillars: the society, living environment and economy. The interdependency and interactions among the principal system components are fully recognized and used in the development of solutions to water problems. The uses of the receiving waters, including natural functions, in-stream uses and withdrawals (e.g. for water supply), are often the driving force dictating the level of control of urban drainage and wastewater effluents.

Recently, the depletion and degradation of urban water resources has led to the advocacy of a sustainable urban water system characterized by lower water consumption, preservation of natural drainage, reduced generation of wastewater through water reuse and recycling, advanced water pollution control, and preservation and/or enhancement of the receiving water ecosystem. Specifically, sustainable urban water systems should fulfil the following basic goals:

- the supply of safe and palatable drinking water to the inhabitants at all times
- collection and treatment of wastewater in order to protect the inhabitants from diseases and the environment from harmful impacts
- control, collection, transport and quality enhancement of stormwater in order to protect the environment and urban areas from flooding and pollution
- reclamation, reuse and recycling of water and nutrients for use in agriculture or households in case of water scarcity.

Further information on sustainability, implementation approaches, and progress indicators can be found in the 2007 UNESCO book on aquatic habitats in this series (Wagner et al., 2007).

Most of the goals of sustainability have been at least partly reached or are within reach in North America and Europe, but are far from being achieved in developing parts of the world, where in 2000 1.1 billion people lacked access to improved water supply services and 2.6 billion lacked adequate sanitation (WHO and UNICEF, 2000).

The Millennium Development Goals put strong emphasis on poverty reduction and reduced child mortality. The two specific goals related to water are:

- to halve, by the year 2015, the proportion of people who are unable to access or afford safe drinking water
- to stop the unsustainable exploitation of water resources by developing water management strategies at local, regional, and national levels which promote both equitable access and adequate supplies.

In addition to these goals, the proportion of people lacking access to adequate sanitation should be halved by 2015. Sanitation systems must be designed to safeguard human health as well as the health of the environment.

The ensuing challenges are exacerbated by the fact that the urban populations of the world, especially in Africa, Asia, and Latin America, are expected to increase dramatically. The African urban population is expected to more than double over the next twenty-five years, while that of Asia will almost double. The urban population of Latin America and the Caribbean is expected to increase by almost 50% over the same period (WHO and UNICEF, 2000). Consequently, the providers of urban services will face great challenges to meet the fast-growing needs over the coming decades. Water management considerations in providing urban water services, including the basic requirements on urban water infrastructure, are presented in this chapter. Three urban infrastructure sub-systems are considered in this discussion: water supply, drainage, and wastewater management and sanitation.

Finally, it is the provision of urban water services and construction of related infrastructure which changes components of the hydrological cycle in urban areas and leads to its replacement by the UWC. Specifically, water supply generally involves the import of large quantities of water into urban areas, sometimes from remote catchments. Some of this water finds its way into urban aquifers, via losses from the water distribution networks. Most of the remaining imported water is used within the urban area and turned into wastewater. Increased catchment imperviousness and hydraulically efficient urban drainage contribute to higher volumes and flow rates of runoff, and reduced recharge of groundwater. The collection of sewage also captures some groundwater through sewer infiltration and thereby reduces groundwater tables in urban areas. For most cities in developed countries, collected wastewater is treated at sewage treatment plants, which discharge their effluents into receiving waters and thereby contribute to higher export of water from the urban area, potentially causing pollution of receiving waters. Thus, elements of the urban water infrastructure and their interactions with the hydrological cycle are of particular interest when dealing with the UWC.

3.2 WATER SUPPLY

The state of urban water services worldwide has been assessed in the *Global Water Supply and Sanitation Assessment 2000 Report* (WHO and UNICEF, 2000). A main finding was that the percentage of people served with some form of improved water supply rose from 79% in 1990 to 82% in 2000. Yet at the beginning of 2000, about one-sixth of the world's population (1.1 billion people) was without access to improved water supply services. The majority of these people lived in Asia and Africa. The situation in rural areas is generally worse than in urban areas, where fast-growing populations create special challenges for improving the urban water services. The urban drinking water supply coverage ranges from 85% in Africa to 100% in Europe and North America. These numbers should not be regarded as reliable, since the definition of urban areas with their large fringe areas is imprecise. The term 'improved water supply' is also fairly generous, comprising not only household connections but also public standpipes, boreholes, protected dug wells, and protected springs and rainwater collection. One of the main barriers to improving water services in developing countries is the high cost of infrastructure.

The consequences of poor water supply and sanitation are severe. One is that large populations continue to live in poverty, which was defined by the Commission of European Communities (2002) as 'not simply … the absence of income and financial resources, but also as encompassing the notion of vulnerability and such factors as access to adequate food supplies, education and health, natural resources and drinking water, land, employment and credit, information and political involvement, services and infrastructure'. The report further concluded that water management is an essential issue in developing policies for poverty reduction in developing countries.

In most cities in North America and Europe and in many cities in other parts of the world, however, the citizens benefit from a water supply service that fulfils the major requirements on water quantity and quality. In these cities, water is normally treated in a water treatment plant and distributed by a pipe system to households. The challenge is to provide similar services to all urban inhabitants in the world, using similar technologies or appropriate alternatives.

The report *Water and Sanitation in the World's Cities: Local Action for Global Goals* (UN-HABITAT, 2003) describes the situation worldwide and gives many illustrative examples from the cities in different parts of the world. A sample of water supply data for the ten most populous cities in the world (in 2000) is presented in Table 3.1. Points

Table 3.1 **Water supply in the world's ten most populous cities (2000)**

City and country	Inhabitants in 2000 (million)	Water supply (m³/s)	Water supply (L/capita/day)
Tokyo, Japan	27.9	81	250
Mexico City, Mexico	19.7	69	331
Sao Paulo, Brazil	17.8	63	306
Shanghai, China	17.2	81	407
New York, USA	16.4	57	300
Mumbai, India	16.4	25	130
Beijing, China	14.2	39	239
Lagos, Nigeria	13.4	(data not available)	(data not available)
Los Angeles, USA	13.1	88	570
Calcutta, India	12.7	25	171

Source: UN-HABITAT, 2003

of interest include (a) a large variation in water supply per capita (130–570 L/capita/day; this rate includes distribution system losses), (b) high withdrawal rates (25–88 m³/s), (c) significant losses (15–56%), and (d) that seven out of the ten cities listed were in developing countries.

Explanations of basic terminology used in water supply, provided by the US Geological Survey, can be found in Table 3.2.

3.2.1 Overview of situation in developing countries

Three factors make water management in developing countries particularly challenging: the current deficit in provision of water supply and sanitation, limited water resources in many developing regions, and rapid population growth. A brief overview of such issues for various developing country regions follows.

3.2.1.1 Middle East and North Africa (MENA)

Renewable water supplies in this region have been a challenge throughout history, and are among the lowest in the world. According to Berkoff (1994), the supplies will decrease to a projected value of 667 m³/capita/y by 2025. More than a third of the available supplies are provided by the water systems outside of the region, which contributes to international disputes. Groundwater represents the main supply and a number of countries in the region continue to mine this resource. Seasonal variations also play an important role and need to be managed. Surface storage of water suffers from large losses by evaporation, so more attention needs to be paid to groundwater storage.

Table 3.2 **Definitions of water use terms**

Term	Definition
Consumptive use	The part of withdrawn water that is evaporated, transpired, incorporated into products or crops, consumed by humans or livestock, or otherwise removed from the immediate water environment.
Conveyance loss	The quantity of water that is lost in transit from a pipe, canal, conduit or ditch by leakage or evaporation.
In-stream use	Water that is used but not withdrawn from a ground or surface-water source, for such purposes as hydroelectric-power generation, navigation, water quality improvement, fish propagation and recreation.
Off-stream use	Water withdrawn or derived from a ground or surface water source for public water supply, industry, irrigation, livestock, thermoelectric-power generation and other uses.
Return flow	The water that reaches a ground or surface water source after release from the point of use, and thus becomes available for further use.
Withdrawal	Water removed from the ground or delivered from a surface–water source for off-stream use.

Source: Mays, 1996.

3.2.1.2 Sub-Saharan Africa (SSA)

Sub-Saharan Africa is a region severely suffering from droughts, with 65% of the rural population and 25% of the urban population lacking adequate water supplies. The consequences of water stress are severe, including famine, diseases and land degradation, and loss of vegetation, domestic stock and wildlife resources. Berkoff (1994) estimated that by 2005, eighteen countries in this region would face water stress which would affect about 600 million people. Droughts also cause severe deterioration of environmental conditions. Groundwater represents the main water supply and is used by three-quarters of the region. Almost all important rivers cross or form international boundaries, contributing to further challenges (Sharma et al., 1996). Lakes and wetlands also represent important water resources for many uses.

3.2.1.3 East, South and South-East Asia (ESSA)

This region is characterized by variability and uncertainty of precipitation patterns, which contributes to seasonal water shortages and flood events. Water shortages are persistent in desert areas of India, China and Pakistan; in more developed areas with low-to-moderate rainfall patterns (e.g. South India and North China), shortages may arise on an annual basis. In higher rainfall areas, as in East India and much of South-East Asia, surpluses occur in the wet season. The water balances in the region are very sensitive to the intensity of the monsoon and other cyclonic storms; weak monsoons can result in significant losses in agriculture and reservoir storage, whereas heavy storms may cause devastating floods (Frederiksen et al., 1993; Prosad et al., 1987). The growth of urban populations creates additional environmental and social challenges (World Bank, 1998). Pollution of rivers, estuaries, coastal waters and groundwater systems with low-to-moderate rainfall patterns has been increasing during the last two decades.

3.2.2 Historical development

Some milestones in the development of modern, centralized drinking water systems are worthwhile to review and present here (WaterWorld and Water and Wastewater International, 2000). As early as 3000 BC, drinking water was distributed in lead and bronze pipes in Greece. In 800 BC the Romans built aqueduct systems that provided water for drinking, street washing and public baths and latrines. In the beginning of the nineteenth century, the first public water supply systems were constructed in North America and in Europe, beginning with Philadelphia in the United States and Paisley in Scotland.

In the middle of the nineteenth century, filter systems were introduced in some cities where the water quality started to be a problem. The spread of cholera made it necessary to use disinfection, and the first chlorination plants were installed around 1900 in Belgium and in New Jersey, USA. During the twentieth century, all large cities in North America and Europe introduced successively more and more advanced treatment of centrally supplied drinking water, including physical, biological and chemical methods. This was primarily done for surface waters; groundwater supplies in many countries even at present either require minimal treatment, or no treatment at all (e.g. in Slovenia and Denmark). Towards the end of the twentieth century, microfiltration of raw drinking water was introduced. This is an innovative treatment process that is being employed on an increasing scale.

Water treatment and distribution is effectively governed by various laws and regulations, such as the US Safe Drinking Water Act (introduced in 1974, amended in 1986 and 1996) and the 1998 European Union Drinking Water Directive 98/83/EC (Council of European Communities, 1998). A broader, internationally developed guidance on drinking water quality can be obtained from the World Health Organization (WHO, 2004).

3.2.3 Water demand

The provision of an adequate water supply and sanitation to the rapidly growing urban population is a problem for government authorities throughout the world. In many developed or developing parts of the world, locating new sources or expanding existing ones is becoming more difficult and costly, and is often physically and economically infeasible. The actual cost of water per cubic metre in second and third-generation water supply projects in some cities has doubled, compared with costs of the first and second-generation projects (Bhatia and Falkenmark, 1993).

The ability to manage existing water resources and plan the development of new ones is tied directly to the ability to assess both current and future water use. Traditionally, the rate of water use per capita or other water-use parameters used in agriculture and industry were considered as 'demands' that had to be adequately covered by managing the supply. Recently, this approach has changed from supply management to demand management, whose main objective is to improve the efficiency and equity of water use and sanitation services. For this purpose, different instruments have been developed and these can be generally classified into the following categories:

- water conservation measures
- economic measures
- information and educational measures
- legal measures.

The efficiency of each of these instruments depends heavily on local conditions. In different sections of this chapter, various aspects of managing water demands for domestic, industrial and agricultural purposes are discussed.

The recent literature on water conservation introduced the 'soft path for water', which can be described as a water management approach based on greatly increased efficiency in end water use, incorporating precise management systems to avoid system losses, and matching system components to the exact quantities and qualities required for appropriate classes and locations of end use (Wolf and Gleick, 2003; Rocky Mountain Institute, 2007). As such, this approach focuses on improving the efficiency and productivity of water use, rather than seeking new water supplies.

Water demand is established by the quantities that consumers use per unit of time (for particular prices of water). Water use can be classified into two basic categories: consumptive and non-consumptive. The former removes water from the immediate water source; in the latter, water is either not diverted from the water sources, or it is diverted and returned immediately to the source at the point of diversion in the same quantity as diverted and at appropriate water quality standards for the source.

Municipal water uses include residential/domestic use (apartments and houses), commercial (stores and small businesses), institutional (hospitals and schools), industrial, and other water uses (fire-fighting, swimming pools and park watering). These uses require water to be withdrawn from surface or groundwater sources. A proportion of the withdrawn water quantity may be returned to the source, often in a different location and time, and at a different quality. Further explanations of individual water uses follow.

Domestic water use includes water used for washing and cooking, flushing toilets, bathing and showering, laundry, house cleaning, yard irrigation, private swimming pools, car washing and other personal uses (e.g. hobbies). Water use by public services includes public swimming pools, institutional uses by government agencies and private firm offices, educational institutions (such as schools, universities and their dormitories), fire-fighting, irrigation of parks and golf courses, health services (hospitals), public hygienic facilities (public baths and toilets), cultural establishments (e.g. libraries and museums), street cleaning and sewer flushing. Other uses include entertainment and sports complexes, the hospitality industry, and barber shops and beauty parlours. Small industries include laundries, workshops and similar establishments. Transportation use includes water used for operation of taxis, buses and other motor transport facilities (stations and garages), ports and airports and railways (stations and workshops).

Good estimates of municipal water demands can be obtained by disaggregating the total delivery of water to urban areas into a number of classes of water use and determining separate average rates of water use for each class. This method is called disaggregate estimation of water use. The disaggregate water uses within some homogenous sectors are less variable than the aggregate water use, and a better accuracy of estimation of water use can therefore be obtained.

In order to estimate the total municipal water use in a city, the study area should first be divided into homogenous sub-areas on the basis of water pressure districts or land use units, and then the water use rates can be assumed to be constant for different users within each sub-area. Temporal (annual, seasonal, monthly etc.) variation should also be considered in disaggregating the water uses.

The most commonly used method for water use forecasting is regression analysis. The independent variables of the regression model should be selected on the basis of the available data for different factors affecting the water use and their relative importance in increasing or decreasing water uses. For example, the most important factor in estimating water use in an urban area is the population of each sub-area. A multiple-regression method can also be used to incorporate more variables correlated with water use in municipal areas to estimate and forecast the water demand in the future. Population, price, income, air temperature and precipitation are some of the variables that have been used by different investigators (Baumann et al., 1998).

Time series analysis has also been used to forecast future water demands. For this purpose, time series of municipal water use and related variables are used to model the historical pattern of variations in water demand. Long memory components, seasonal and non-seasonal variations, jumps and outlier data should be carefully identified and used in modelling water demand time series.

In recent years, more attention has been given to conserving water rather than developing new water sources. In many countries, this has been accepted for both economic and environmental reasons as the best solution for meeting future water demands. Water demand management is an appropriate strategy to improve efficiency and sustainable use of water resources, taking into account economic, social, and environmental considerations (Wegelin-Schuringa, 1999; Butler and Memon, 2005). In this approach, increasing attention is paid to water losses and unaccounted for water, which generally include:

- leakage from pipes, valves, meters, etc.
- leakage/losses from reservoirs (including evaporation and overflows)
- water used in the treatment process (back-wash, cooling, pumping, etc.) or for flushing pipes and reservoirs.

Typically, the losses may vary from 10 to 60% (See Table 3.1), with the higher values reported for developing countries. Other measures focus on water saving technologies, such as dual-flush toilets, flow restrictors on showers and automatic flush controllers for public urinals, automatic timers on fixed garden sprinklers, moisture sensors in public gardens, and improved leakage control in domestic and municipal distribution systems. All these measures are practical, but regulations and incentives are needed for their implementation.

The costs of water supply services – and technological developments designed to lower these costs – have a major influence on the level of water demand in the developing countries. Design of a rational water tariff structure is very difficult and challenging, as it must attempt to balance conflicting objectives, including affordability, economic efficiency, equity and cost recovery (Raghupati and Foster, 2002). Raghavendra (2006) examined the assertion that Indian water utilities charge low water tariffs, and concluded that the tariffs charged are not low when considering the quality of services provided. Furthermore, a major challenge in almost all cities in developing countries is that there is a mixture of measured and unmeasured tariffs due to relatively low meter coverage. Furthermore, the tariff structure can be very complex and non-uniform, with tariffs based on metered block consumption, flat rates, ferrule-based charges, annual rentable value and, occasionally, tap-based charges (Raghupati and Foster, 2002). In rural areas, the major factors influencing demand are the distance from households to the standpipes, or the number of persons served by a single tap or well.

3.2.3.1 Water supply standards: quantity

Adequate quantities of water for meeting basic human needs are a prerequisite for human existence, health and development. If development is to be sustained, an adequate quantity of water must be available. In fact, as development increases, in most instances the demand for water on a per capita basis will also increase for personal, commercial, industrial and agricultural purposes.

Per capita water consumption can be measured (or estimated) through metered supply, local surveys, sample surveys or the total amount supplied to a community divided by the number of inhabitants. In planning, the per capita water use is usually controlled by locally specific regulations or standards. The actual domestic water use rates vary widely, from a minimum of 50 L/capita/day (Gleick, 1998) to 500 L/capita/day (or more), depending on water availability, pricing, traditional water use and other factors. In urban areas with high levels of water use, there is scope for creating large reserves by introducing and practising demand side management (Baumann et al., 1998).

The supply source and the storage facilities in urban areas are designed to yield enough water to meet both the current daily demands and the consumption forecasted in the near future. Water supply systems in urban areas should satisfy quantitative guidelines and standards. As a general rule, when using surface supplies, the tributary watershed should yield the estimated maximum daily demand for ten years into the future, and the storage capacity of a supply reservoir should be equal to at least thirty days maximum daily demand five years into the future. Ideally, for well supplies there should be no mining of water: that is, neither the static groundwater level nor the specific capacity of the wells (litres per minute per metre of drawdown) should decrease appreciably as demand increases. These values should be constant over a period of five years except for minor variations that correct themselves within one week.

3.2.3.2 Water supply standards: quality

The quality of water is assessed in terms of its physical, chemical and biological characteristics and its intended uses. Water to be used for public water supplies must be potable (drinkable), that is without polluting contaminants that would degrade the water quality and constitute a hazard or impair the usefulness of the water. To ensure drinking water safety, the multiple barrier approach is advocated; this consists of an integrated system of measures that safeguard water quality from the source to the tap. For safety, a redundancy of protection measures is built into these systems. The quality of drinking water is prescribed by the appropriate standards (WHO, 2004). In this context, the term 'standard' refers to a definite rule, principle or measure established by an authority. Where health is of concern and scientific data are limited, precautionary standards may be justified.

Quality criteria for drinking water have been presented in many documents. Particularly well known are the regulations mandated by the US Environmental Protection Agency, title 40, parts 141 and 143 and the Safe Drinking Water Act (1974, amended in 1986 and 1996). These are the current regulations for evaluating the suitability of surface or groundwater resources for public water supply in the United States (for the latest version, visit www.epa.gov/safewater/mcl.html), and provide primary and secondary standards. The primary standard is for human health protection; the secondary standard underpins a regulation that specifies the maximum contamination levels that are permissible without risk to the public welfare, but that may adversely affect the appearance or odour of water.

In the international forum, the most authoritative document on drinking water quality is the World Health Organization's *Guidelines for Drinking-Water Quality* (WHO, 2004). These guidelines address a framework for safe drinking water, health-based targets, water safety plans, surveillance, applications of guidelines in specific circumstances, microbial aspects, chemical aspects, radiological aspects and acceptability aspects. Specific circumstances include emergencies and disasters, large buildings, packaged/bottled water, travellers, desalination systems, food production and processing, and water safety on ships and in aviation. The guidelines are suitable for use by both developed and developing countries. The most frequent concerns about drinking water quality are those posed by pathogens and arsenic.

Requirements for the quality of industrial water supply depend on the type of industry, and may even differ in various segments of a particular industrial sector. For a

detailed description of water quality requirements in various types of industries, the reader is referred to Corbitt (1990).

There are many water needs that can be met by sub-potable water. Typical examples are irrigation of urban landscape, agricultural irrigation, aquaculture, some domestic uses (e.g. toilet flushing), industrial reuse (e.g. cooling waters or process waters), recreational waters and groundwater recharge. Sub-potable water requirements can be met by reclaimed urban wastewater, with the main benefits consisting in saving potable water and reducing pollution discharges into receiving waters. In water reuse, the most important issue is to specify the quality of water to be reused for a particular purpose. Such specifications then determine the required level of treatment. Some guidance for water reuse can be obtained from the WHO guidelines (WHO, 1989), which are under review. Further discussion of wastewater reclamation and reuse is presented in Section 3.4.10.

3.2.4 Water supply sources

The gap between society's needs for water and the capability to meet such needs is continually widening. Water supply, an important element of the overall UWC, attempts to bridge this gap. Water needed in urban areas may come from groundwater or from surface water sources such as lakes, reservoirs and rivers. It is called untreated or raw water, and is usually transported to a water treatment plant. The degree of treatment depends on the raw water quality and the purpose this water will be used for. Different water quality standards for municipal purposes have been developed and used during the past several decades. After treatment, the water is usually distributed via a water distribution network.

Four major characteristics of water supply are quantity, quality, time variation and price. If the quantity and time distribution of raw water conformed to the water use patterns in an urban area, then there would be no need to store water or regulate its distribution by artificial structures or devices. But in almost all urban areas, the time variation of available water resources does not follow demand variations. Therefore, certain facilities need to be created to store the excess water during wet (high-flow) seasons so that it can be consumed in low-flow periods.

The costs of initial investment in, and operation and maintenance of, the water supply should be incorporated in the economic studies for development planning. In the same way, if water quality does not satisfy the standards for different water uses, treatment plants should be set up and their costs should be incorporated in the urban water resources development studies. In general, water supply methods can be classified as:

- conventional methods of surface and groundwater resources development
- non-conventional methods.

Conventional methods of water supply include large-scale facilities such as dam reservoirs, water transfer structures and well fields. Dam reservoirs are the most important sources of water in many large cities around the world. Quite often the reservoirs providing the water supplies of urban areas are located tens or hundreds of kilometres away from the areas served, sometimes in another river basin.

Water supply storage facilities range from large reservoirs created by building dams to small-scale storage tanks. The term 'reservoir' has a specific meaning with regard to water supply systems modelling and operation. It is an 'infinite' source that can supply or accept water and has such a large capacity that the hydraulic grade elevation of the reservoir is unaffected and remains constant.

In many large cities around the world, water demand has exceeded the total water resources in the basin in which the city is located. In such cases, one of the approaches to water supply management is to develop new inter-basin water transfer schemes in order to keep ahead of the ever increasing requirements of the growing population and improved standards of living. A different approach is referred to as demand side management, in which water conservation is practised to the maximum practical extent to reduce demands on the water supply and manage the needs for infrastructure expansion (Butler and Memon, 2005).

3.2.4.1 Conjunctive use of sources and artificial recharge

Conjunctive use of surface water and groundwater can sometimes offer an attractive and economical means of solving water supply problems where the exploitation of either resource is approaching or exceeding its optimum yield. The volumes of groundwater naturally replaced each year are relatively small because of the slow rates of groundwater movement and the limited rate of infiltration. However, artificial recharge can be used to reduce adverse groundwater conditions such as progressive lowering of water levels or saline water intrusion. As shown in Table 3.3, the percentage of population in selected countries relying on groundwater supply ranges from 15 to 98%; the rest is served by surface water sources.

3.2.4.2. Supplementary sources of water

Three supplementary sources of water are of particular importance in urban areas: rainwater harvesting, bottled water and wastewater reclamation and reuse. The first source is addressed in this section, the second in Section 3.2.8, and the third in Section 3.4.10.

Table 3.3 **Percentage of the population supplied by groundwater in selected countries**

Country	Population supplied by groundwater (%)
Denmark	98
Portugal	94
Italy	89
Mexico	75
Switzerland	75
Belgium	67
Netherlands	67
Luxemburg	66
Sweden	49
United States	40
United Kingdom	35
Canada	25
Spain	20
Norway	15

Rainwater harvesting is a supplementary or even primary water source at the household or small community level, especially in places with relatively high rainfall and limited surface waters (e.g. small islands) and in developing countries.

The potential of domestic rainwater harvesting to improve urban slum dwellers access to water in Africa has been assessed by Cowden et al. (2006). They found that on an annual basis, 68% of urban slum dwellers would potentially be able to obtain at least three months of water supply at 20 L/capita/day, and 23% of slum dwellers would be able to obtain at least three months of water supply at 50 L/capita/day using rainwater harvesting technology.

The use of roof runoff for irrigation or other uses is also becoming of interest in highly developed urban areas. It is one of the measures that can support environmental sustainability by reducing water supply demands for irrigation and lessening urban runoff and its impacts. Roofs of buildings are the most common collecting surfaces, but natural and artificial ground collectors are also used in different places around the world. In designing rainfall harvesting systems two kinds of issue should be considered:

- *Quantity issues*: Rainwater collection systems often suffer from inadequate storage tank volumes or collector areas. Leakage from tanks due to poor design, selection of materials or construction – or a combination of these factors – is a major problem of rainwater collection systems.
- *Quality issues*: The rainwater quality in many parts of the world is good. But water quality problems may arise within the collection systems. Physical, chemical and biological pollution of rainwater collection systems occurs where inappropriate construction materials have been used or where maintenance of roofs and other catchment surfaces, gutters, pipes and tanks is not properly performed (Falkland, 1991). The feasibility of disinfecting harvested rainwater was examined by Zhu et al. (2006).

Rainwater cisterns or other rainwater collection devices have been used in many parts of the world for centuries. At present, this approach is used widely in Australia and India, and is gaining acceptance in China (Zhu et al., 2006; Li and Geiger, 2006). In rural areas of Australia, many farms have access to neither a water supply system nor adequate well water; hence, the rainwater tank has become a symbol of the Australian outback culture. Local councils and urban water authorities are increasingly encouraging these systems in urban areas, and in some cases offering rebates to customers who use rainwater for sub-potable uses. The most feasible reuse of rainwater in urban areas is for garden irrigation, which accounts for 35 to 50% of domestic water use in many large cities of the world. Reuse of rainwater in the garden requires a relatively simple system, with very low environmental risks, and it is therefore encouraged by many water authorities. Examples of rooftop rainwater harvesting systems have been presented by Karamouz et al. (2003) elsewhere.

Further savings of potable water can be achieved when rainwater is used for toilet flushing (about 20% of domestic water use), as well as in the laundry, kitchen and bathroom. It can also be used in pools, and for washing cars. In some situations (e.g. in some rural areas), it may be possible to use rainwater for most domestic uses, without relying on the public water supply. In all these cases, strict regulations for reclaimed water quality must be followed and safety systems employed, particularly in connection with drinking water, which should be protected by a 'multiple barrier system'. In this approach, multiple barriers are used to control microbiological pathogens and contaminants that may enter the water supply system, thereby ensuring clean, safe and reliable drinking water.

3.2.4.3 Water shortage

Urban demands on water supplies are continually increasing as a result of growing urban populations and higher standards of living. When demands exceed the available water, shortages result, with significant social, political and economic implications. In general, there are two main reasons for water shortage in urban areas: (a) climatic or hydrologic drought, and (b) the inability of the supplier to provide the required water.

Social impacts of water shortage are related to the effects of water deficits on public health and lifestyles; such impacts are exacerbated when shortages cause inequitable water distribution. As a part of their long-term planning, urban water suppliers need to ensure an appropriate level of reliability of water supply that will meet the needs of various categories of customers during both normal and dry years.

The capacity of a water supply system to meet the customer demands is measured by different performance parameters, such as reliability (reflecting the probability that the system will meet demands), resiliency (the speed of returning the system to a fully operational state), and vulnerability (reflecting the consequences of the failure) (Hashimoto et al., 1982).

In comprehensive water supply planning, the following factors are considered important to mitigate water shortages:

- planning horizon
- planning criteria (including reliability, cost, and water quality)
- demand projections
- water availability under the current conditions (the minimum water supply is estimated for the driest water year in the planning horizon, all supply opportunities such as recycled water and water transfer are considered, and plans for replacing water resources at risk are assessed)
- a long-term water supply strategy
- water quality considerations
- treatment and production facilities
- contingency plans (may include water use restrictions and rationing, prioritizing competing uses, and utilizing alternative sources).

Water shortages may lead to conflicts among the stakeholders, which need to be resolved.

3.2.5 Example of improving the water supply in the Tehran Metropolitan Area

Water supply shortcomings can be sometimes alleviated by water transfer, in this case from south-east to south-west of Tehran. The combination of a shortage of water and the availability of more suitable and cultivated lands in the western area in comparison with the eastern part led Karamouz et al. (2005a) to propose a water-transfer channel from east to west. This channel crosses all the local rivers and wastewater channels from east to west in the study area before reaching the irrigation lands. Its construction has made it possible to transfer up to $7\,m^3/s$ of mostly urban drainage water from east to west in the region. This has helped to overcome the negative impacts of water rise in the east, and to utilize the urban drainage water in the cultivated lands and for other municipal applications in the west. The channel intercepts several local rivers and drainage channels and could collect some of the water from these outlets and supply it to farmers, who prefer to use surface water for irrigation, rather than more expensive groundwater withdrawals.

The surface water was allocated to each geographical zone on the basis of a weighting system related to the groundwater table in the zone and demand. The relative weighting given to each crop was based on its price. In order to simulate the state of the system under different water transfer scenarios and water allocation schemes, an object-oriented simulation model has been developed (Karamouz et al., 2005b). Using different objects, the user can change the allocation schemes for any given demand points. An economic model based on conflict resolution methods was also developed for resource planning in the study area.

3.2.6 Drinking water treatment

Treatment is generally required to make raw water drinkable. As high-quality sources of water are depleted, water utilities are increasingly using lower-quality source water that requires more treatment and consumes more water during the treatment. Detailed discussion of water treatment is beyond the scope of this report and can be found elsewhere (Pontius, 1990). Instead, the material presented here focuses on emerging technologies, desalination and disinfection.

3.2.6.1 Emerging technologies

Rapid development of microfiltration (membrane filtration) has created new treatment options that were not feasible previously. One such option is a system that delivers raw source water, or water that has had only primary treatment, directly to the user, who further treats this water in small, local treatment units near the point of consumption. The main advantages of this system are that the deterioration of the water quality during transport is avoided and different consumers can treat the water to the level specifically needed for their requirements. Needless to say, the water quality does not need to be the same for such purposes as human consumption (drinking water), process water for industry, cooling water, irrigation water or water for flushing toilets. Small, locally used microfiltration units are rapidly becoming competitive in price. The conventional central and alternative on-site water treatment systems are illustrated in Figures 3.2 and 3.3.

3.2.6.2 Desalination

Among unconventional water supply methods, the use of seawater treated by desalination is the most widely used. Seawater has a salinity of about 35,000 mg/L, which is attributable mostly to sodium chloride. Desalination was first adopted in the early

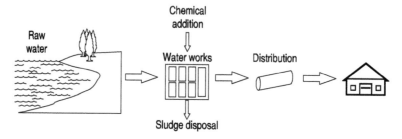

Figure 3.2 **A conventional layout of water treatment and distribution systems**

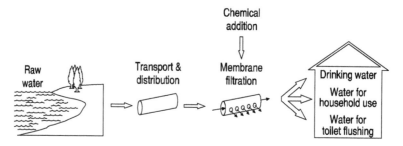

Figure 3.3 **An alternative layout of drinking water distribution and treatment systems using small, local membrane filter units**

1960s, using Multi-Stage Flash evaporation (MSF), and a number of plants using various desalination technologies are currently operating in the United States, the Caribbean and the Middle East. Desalination technologies have increasingly become cheaper and more reliable, and their current costs may be competitive with those of other high-tech treatment technologies, especially if water sources are sparse or very remote.

The following methods are used for removing dissolved solids:

- *Distillation*: In this method, the water is heated to its boiling point to convert it into steam; the steam is then condensed, yielding salt-free water.
- *Reverse Osmosis (RO)*: In this method, the water is forced through a semi-permeable membrane under pressure; the dissolved solids are held back.
- *Electrodialysis*: In this method, ions are separated from the water by attraction through selective ion-permeable membranes using an electrical potential.

Among the above methods, the first two are more common for desalination of seawater; electrodialysis is usually preferred for treating brackish groundwater. During the last decade the development of membrane technologies has been rapid and has resulted in the development of new, cheaper and high-performance membranes. Plants using membrane technologies (RO and electrodialysis) have been built and the cost of the finished water is steadily decreasing. The main advantage of membrane technologies over distillation methods is that they use much less energy. Pre-filtration is often used to remove hydrocarbons and reduce the fouling of the RO membranes. The cost of RO desalination has decreased substantially over the past decade, falling to between

US$0.25 and US$1.00 per cubic metre. Even so, seawater desalination remains more expensive than most other sources of water supply (where available), but in arid locations close to the sea and far from suitable surface or groundwater sources, seawater desalination may be the most economical option for an urban water supply. At a lower level, small-scale membrane units have been developed and sold in all parts of the world to offices and single households. They are becoming common amongst the middle-class households in Delhi (India) and other cities of India where the water supplied has turned brackish due to excessive groundwater withdrawal. The cost of such units (typically around US$2,000 per unit) is decreasing but still too high for common use. As the technology improves further (particularly for reverse osmosis), these costs may decrease.

In addition to high rates of energy consumption, a major problem at inland desalination plants is the disposal of rejected brine. Evaporation ponds, injection in deep wells, and transfer to the ocean are the common methods of dealing with this, depending on the volume of brine, site location and geographical and climatic conditions.

3.2.6.3 Disinfection

The introduction of chlorine to disinfect drinking water in the early twentieth century drastically reduced waterborne diseases in western cities. Even today, the struggle to eliminate pathogens and deliver a healthy municipal drinking water is of primary concern all over the world. In remote or poor regions of the world, boiling of drinking water will continue to be used for a long time to protect against diseases. Alternative disinfection methods have been developed and used, most often by chlorination using hypochlorite, chlorine dioxide and chloramines.

One major drawback of adding chlorine is the formation of carcinogenic by-products during the disinfection process. Among such products, the most common are trihalomethanes (THM), haloacetic acids, bromate and chlorite. Numerous studies have demonstrated that these byproducts may increase cancer risks (e.g. Batterman et al., 2002; Gibbons and Laha, 1999; Goldman and Murr, 2002; Korn et al., 2002). Consequently, the US Environmental Protection Agency (EPA) limits the presence of THM in drinking water to 80 μg/L (Gibbons and Laha, 1999). When considering the risks associated with these substances, one should keep in mind that such risks are much smaller than the microbiological risks incurred if water is not disinfected. To avoid the problems associated with chlorination byproducts, other disinfection methods have been introduced, including ultraviolet (UV) irradiation, ozonation and solar disinfection. In developing countries, the first two methods may not be affordable, but the third has a great potential, because of its simplicity and low costs.

UV irradiation and ozonation may require pre-treatment and, before discharging disinfected water into the distribution system, some chlorine may have to be added to the treated water. This may be needed to maintain chlorine residuals during transport in the distribution network and prevent growth of bacteria originating from biofilms found in the pipe system. In many countries the residual chlorine concentration at the tap is taken as a measure of safety.

In solar disinfection, solar radiation is used to destroy pathogenic micro-organisms which cause waterborne diseases. This process is best suited for treating small quantities of water, usually placed in transparent plastic bottles and exposed to full sunlight for extended time periods. The actual treatment occurs through solar UV irradiation and increased water temperature, ideally above 50°C, which accelerates the death of the bacteria.

3.2.7 Water distribution systems

Water distribution systems comprise the entire infrastructure from the treatment plant outlet to the tap. The actual layout of supply mains, arteries and secondary distribution feeders should be designed to deliver, above and beyond the maximum daily rate usage, sufficient capacity for fire-fighting in all built-up parts of the municipality. Similar considerations must be given to the effects that a break, joint separation, or other mains failure could have on the water distribution system. In urban areas, most of the water quantity standards relate to the distribution system, as further discussed below, on the basis of general practices in this field.

In evaluating a water distribution system, pumps should be considered at their effective capacities when discharging at normal operating pressures. The pumping capacity, in conjunction with storage, should be sufficient to maintain the maximum daily use rate plus the maximum required fire-fighting flow with the single most important pump out of service.

Storage is frequently used to equalize pumping rates in the distribution system as well as to provide water for fire-fighting. In determining the fire flow from storage, it is necessary to calculate the rate of delivery during the specified period. Even though the volume stored may be large, the flow to a hydrant cannot exceed the carrying capacity of the mains, and the residual pressure at the point of use should not be less than 140 kPa.

Depending on specific practices, the recommended water pressure in a distribution system is typically 450 to 520 kPa, which is considered adequate for buildings up to ten stories high as well as for automatic sprinkler systems for fire protection in buildings of four to five stories. For a residential service connection, the minimum pressure in the water distribution main should be 280 kPa; pressure in excess of 700 kPa is not desirable, and the maximum allowable pressure is 1030 kPa. Fire hydrants are installed at spacing of 90–240 m in locations required for fire-fighting.

Although a gravity system delivering water without the use of pumps is desirable from a fire protection standpoint, well-designed and properly safeguarded pumping systems can be made so reliable that no distinction is made between the reliability of gravity-fed and pump-fed systems. Electric power should be provided to all pumping stations and treatment facilities by two separate lines from different sources. For more detailed information about pumps and the quantitative standards in water distribution systems, the reader is referred to the water distribution textbooks such as Rishel (2002) and Walski et al. (2003), respectively. The use of standby emergency power is also recommended.

3.2.8 Drinking water supplies in developing countries

It is estimated that over a third of the urban water supply systems in Africa, Latin America and Asia operate only intermittently. An intermittent water supply is a significant constraint on the availability of water for hygiene, and encourages the low-income urban population to turn to alternatives such as water vendors.

Many of the intermittently operating systems do not deliver water more than half the time, and there are large variations in water quality. One way of overcoming the lack of water supply is to construct local water reservoirs, from which water is delivered to outdoor taps or into the households. Problems arise from the facts that these reservoirs are poorly protected against tampering and seldom cleaned. Further, the variations of pressure in the pipe systems may cause intrusion of contaminated water. Considerable risks of spreading diseases exist in such systems (WHO and UNICEF, 2000).

Among the less-developed but commonly used technologies are unprotected wells and springs, vendor-provided water, bottled water and tanker truck provision.

Vendor-provided water is rapidly expanding in many countries and raises many questions concerning water quality and price. It is often argued that for the price of the vendor-provided drinking water, more efficient communal water supply systems could be constructed. However, water vendors today play an important role in many regions of the world.

The bottled water industry is growing quickly all over the world and serves an important market. In developed countries, the price of bottled water is about 1,000 times higher than that of the public water delivered at the tap. Nevertheless the market is growing, influenced by 'trendy' consumers and by people who do not trust the quality of the tap water. In developing countries, bottled water, where affordably priced, may solve acute drinking water problems or problems with low-quality water.

Tanker truck provision is common in many regions where there is no public distribution system. Typical examples may be small villages or peri-urban areas. Generally (e.g. in India), this water is supplied from public water supplies by contractors.

3.3 URBAN DRAINAGE

Urban drainage serves to reduce the risk of flooding and inconvenience due to surface water ponding, alleviate health hazards, and improve aesthetics of urban areas. Traditionally, drainage development was based on a steady expansion of the drainage infrastructure, and there was no consideration of the impacts of drainage discharges on receiving waters and their beneficial uses.

Two interconnected urban drainage systems are recognized: the major system serving to alleviate major flooding, and the minor system providing convenience by reducing water ponding in urban areas. The minor system comprises swales, gutters, stormwater sewers, open drains, and surface and subsurface storage facilities, and conveys runoff from frequent events with return periods of up to ten years. This system reduces the frequency of inconvenience (water ponding) and its failure has minimal implications. The major system consists of natural streams and valleys, as well as large constructed drainage elements such as large swales, streets, channels and ponds. The major drainage system greatly reduces the risk of fatalities and property damage in urban areas, and consequently its failure has serious consequences. It is typically designed for a 50-year or even a 100-year event (Geiger et al., 1987).

There are two types of urban sewerage systems: combined and separate. The combined system conveys both surface runoff and municipal wastewaters in a single pipe. In dry weather, the entire flow is transported to the sewage treatment plant and treated. In wet weather, as the runoff inflow into the combined sewers increases, the capacity of the collection system is exceeded and the excess flows are allowed to escape from the collection system into the receiving waters in the form of the combined sewer overflows (CSOs), which pollute the receiving waters.

In the separate system, surface runoff is transported by storm sewers and discharged, with or without passive treatment, into the receiving waters, and the municipal wastewater is transported by sanitary sewers to the wastewater treatment plant (WWTP) and usually treated prior to discharge into the receiving waters. Both drainage systems have evolved throughout history and currently exist in many variations. Existing systems represent a compromise among providing flood protection, improving quality of life, technical abilities, ecological needs and the availability of resources.

Urban drainage interacts with other components of the UWC, and particularly with receiving waters. Fast runoff from impervious surfaces, together with hydraulic improvements of urban drainage in the form of street gutters, storm sewers and drains, results in an increased incidence and magnitude of stormwater runoff. The resulting high flows affect the flow regime, sediment regime, habitat conditions and biota in receiving waters. Urban drainage can also result in low flows. Reduced infiltration leads to reduced groundwater recharge, lowered groundwater tables and reduced base flows in rivers. Low flows reduce the self-purification capacity of rivers and limit the dilution of polluted influents, and consequently are characterized by poor water quality. Where groundwater is withdrawn for urban water supply, aquifers are often over-exploited and land subsidence may occur.

Drainage also interacts with other water infrastructures and water resources. For example, cross-connections between storm and sanitary sewers either allow municipal sewage to flow into separate storm sewers with concomitant pollution of stormwater, or stormwater to flow into sanitary sewers and increase the flow rates, which may exceed the WWTP capacity, resulting in sewage bypasses, and the pollution of receiving waters. All types of sewers may interact with groundwater, particularly if not watertight. Leaky sewers below the water table suffer from groundwater infiltration and essentially drain groundwater aquifers. Infiltration of groundwater into sanitary sewers increases sewage flows reaching the WWTP and thereby increases the cost of treatment and the risk of sanitary sewer overflows. On the other hand, exfiltration from sanitary sewers pollutes groundwater in urban areas and may impact on sources of drinking water. Sudden increases in wet weather flows in combined sewers produce hydraulic and pollution shocks on the treatment plants and may reduce the treatment efficiency, particularly of biological treatment by shortening the reaction time and reducing the return sludge flow. In addition, the biomass is diminished as sludge is flushed into the final clarifier. All these factors can lead to reduced treatment efficiencies and increased discharge of pollutants into the receiving waters.

In developing countries, the urban poor inhabiting marginal land prone to flooding and landslides are affected the most by poor drainage (Parkinson, 2002). In many such areas, urban drainage is still designed for quick removal of water from urban surfaces, with concomitant increases of the peak flows and the costs of drainage systems. These impacts are reflected in the operation of the major drainage (Tucci, 1991).

Furthermore, in poorly drained areas urban stormwater mixes with sewage from overflowing latrines and sewers, and causes pollution and spread of waterborne diseases. Other health problems are caused by flooded septic tanks and leach pits (breeding of mosquitoes), and faecally contaminated wet soils contribute to the spread of intestinal worm infections. Contaminated water may gain entry into low (or negative) pressure water-distribution systems operating intermittently, contaminate drinking water, and cause outbreaks of waterborne diseases (Parkinson, 2002).

Many problems associated with the operation of stormwater drainage systems, particularly those comprising drainage ditches, are linked to poor solid waste management. Drainage ditches are often used illicitly for solid waste disposal and municipal agencies generally lack sufficient resources and equipment for drain cleaning. Other problems are caused by illegal construction in floodplains (Parkinson, 2002).

Similar findings were reported by Silveira (2002), who reported that socio-economic factors encountered in developing countries make it challenging to solve problems of urban drainage. The barriers to adoption of modern solutions include:

- the prevalence of the old nineteenth-century sanitary philosophy
- both legal and illicit land settlements that limit the land available for implementing modern drainage solutions
- contamination of stormwater by sanitary sewage, sediment and solid waste prevents adoption of modern drainage practices
- climatic and socio-economic factors that contribute to disease spreading where runoff is stored for flood control or to enhance infiltration
- lack of technology for drainage management and design
- the lack of interaction and collaboration between the community and city administration needed to develop modern solutions.

Concerning drainage flows, urban water managers are interested in both water quantity and water quality issues, as discussed in the following sections.

3.3.1 Flooding in urban areas

Floods are naturally occurring hydrological events characterized by high discharges and/or water levels leading to inundation of land adjacent to streams, rivers, lakes or coastal areas. Where such areas are occupied by human settlements, disasters may occur and result in loss of human life and material damage. Two types of floods are distinguished in urban areas: those locally generated by high intensity rainfall, and those generated in larger river catchments and passing through urban areas, where they may inundate flood plains which have been encroached upon. Other floods may occur in coastal areas, in the form of storm surges or tsunamis with catastrophic impacts. Only the first flood type, locally generated, will be addressed here; river floods or flooding of coastal areas are beyond the scope of this report.

Locally generated floods usually result from catchment urbanization, as discussed in Chapter 4. High catchment imperviousness, hydraulically efficient flow conveyance, and reduced concentration times causing runoff generation by high-intensity rainfall all contribute to high rates of surface runoff and risk of local flooding. In developed countries, these issues have been addressed with various degree of success by applying both non-structural and structural measures incorporated in a master drainage plan. The methodology for urban drainage planning is well developed and described in Geiger et al. (1987). These plans are prepared at two planning levels, short-term (five to ten years) and long-term (twenty-five to fifty years). A master drainage plan represents a technical layout of the sewerage systems (drainage and sanitation) for the entire urban area as it may further develop within the planning horizon. With respect to drainage, the master plan should be part of a catchment plan and incorporate the whole drainage system, including the connections and interactions between the minor and major system components.

Typical master drainage plans include such points as purpose and background of the study, identification of drainage-related problems, definition of study objectives, a database for planning, methods for planning and design, identification and investigation of drainage alternatives, the impact of the future drainage system, final design of individual structures, and implementation. Much success has been achieved with master drainage planning in developed countries, where this planning process is well established. In developing countries, progress in drainage planning is being made, as documented by publication of guidelines for modern drainage design (Indian Road Congress, 1999; Department of Irrigation and Drainage, 2000).

In some developing countries, urbanization occurs too fast and unpredictably, and often progresses from downstream to upstream areas, which increases flood problems (Dunne, 1986). Urbanization of peri-urban areas is largely unregulated, often without the provision of any infrastructure, many public lands are occupied and developed illegally, and flood-risk areas (flood plains) are occupied by low-income populations without any protection. WHO (1988) reported spontaneous housing developments in flood-prone areas of many cities in humid tropics, including Bangkok, Mumbai, Guayaquil, Lagos, Monrovia, Port Moresby and Recife.

Other problems include lack of funding for drainage and other services, lack of solid waste collection (which may end up in and block drainage ditches), no prevention of occupation of flood-prone areas, lack of knowledge about coping with floods, and lack of institutions in charge of flood protection and drainage (Dunne, 1986; Ruiter, 1990). Tucci and Villanueva (2004) suggested solutions, including the introduction of better drainage policies (which would control flow volumes and peaks), and planned development, in which space is retained for flow management measures. Also, in flood plains, non-structural measures should be applied by emphasizing green areas, paying for relocation from flood-prone areas, and public education about floods.

3.3.1.1 Local (pluvial) flooding

The main concerns when sewers flood in urban areas are economic losses and public health issues. The total economic loss includes damage to buildings and their contents and to the infrastructure, and disruption of business activities. However, the total effect of pluvial flooding is difficult to calculate.

Traditionally, indicators of the risk associated with flooded sewers were oriented towards the hazard. Decades ago rainfall intensity–duration–frequency curves generated from historical recordings were used as hazard indicators (Marsalek et al., 1993), but the current computational methods provide a better indicator: water levels. As an indicator, the water level has the advantage of a much closer connection to damage. As long as the water stays below the street level, flood damage is limited, but it starts to rise when water starts to accumulate on the street level. Thus, water level and rain intensity are both indicators of the hazards related to pluvial flooding; the second of these is a natural phenomenon which can not be controlled. Other risk reduction measures need to be applied, such as early warning (which serves to reduce damage), construction of sewer systems with reduced vulnerability, and reducing inflow into sewers by means of source controls or other stormwater management measures (Hauger et al., 2006).

At present, the most important reason, and the one with the most significant economic consequences, for updating the systems in many parts of the world is climate change. In some locations, the expectation is that annual precipitation will increase slightly and will concentrate in fewer but more intense events (Ashley et al., 2004; Grum et al., 2006). A hazard-focused strategy would increase sewer pipe diameters and incorporate large storage facilities, with some economic consequences. A different approach advocated by urban planners focuses on reducing vulnerability by controlling the volumes of stormwater and directing them to the areas where flooding will cause limited damage.

3.3.1.2 Urban (fluvial) floods

Urban areas also have to cope with fluvial floods, which cause about 40% of all deaths caused by natural disasters, mostly in the developing countries. For example, 3.7 million people were killed in a 1931 flood on China's Yangtze River. Floods also bring about environmental and social benefits, in the form of supply of sediments and nutrients to floodplains and downstream areas. This process is important to river ecology and for agricultural production. Many farmers in the developing countries plant crops in soils saturated by receding floodwaters. Floods also support fisheries and food security for the people relying on such a food supply. Thus, flood management needs to balance flood benefits and damage.

Poor land use policies and weak enforcement contribute to development encroachment onto floodplains, particularly during periods without large floods. When higher floods return, damage increases and the public administrations are forced to invest in flood relief. Structural solutions can be expensive and are feasible only when preventing even greater damage, or for intangible social considerations. Non-structural measures are less expensive, but are generally not politically attractive (Tucci, 1991).

The essential principles of urban flood management were summarized by Tucci (1991) as follows:

- Flood control evaluation should be done for the whole catchment, rather than just specific sections.
- Urban drainage control scenarios should take account of the future city development.
- Flood control measures should avoid transferring flood impacts to downstream reaches, by giving priority to source control measures.
- The water quality impacts of urban runoff should be controlled.
- Flood management plans should emphasize non-structural measures for flood management, such as flood mapping, zoning, insurance and real-time flood forecasting and warning.
- Management of flood control starts with the implementation of an urban drainage master plan in the municipality.
- Public participation in the urban drainage management should be increased.
- The development of urban drainage should be based on the cost recovery of investments.

While these principles have been applied in many developed countries, the drainage practice in most developing countries falls short of applying these principles for the following reasons:

- Urban development in the developing countries is too fast and unpredictable, often progressing from the downstream areas towards the upstream areas, which increases the damage impacts (Dunne, 1986).
- Urbanization of undeveloped areas usually ignores municipal regulations for land development, for reasons explained below.
- Unregulated developments – on undeveloped land on the periphery of large cities – may be relatively inexpensive and it is advantageous for private landowners to develop such land without infrastructure and sell it to the low-income population.
- Invasion of public areas (such as public green areas, designated in the urban master plan) by squatters. Due to poverty and slow decision making by the public administration, these developments may become consolidated without proper water and electricity services. In 1973, 59% of the city of Bogota comprised illegal developments and most of them lacked sewers, water and electricity;
- Occupation of peri-urban and risk-prone areas without any infrastructure (floodplains and hill-side slope areas) by low-income populations. Spontaneous housing developments in risk-prone areas in humid tropics cities are often on land prone to flooding (Bangkok, Bombay, Guayaquil, Lagos, Monrovia, Port Moresby and Recife), or on hillsides prone to landslides (Caracas, Guatemala City, La Paz, Rio de Janeiro and El Salvador) (WHO, 1988).
- Municipalities and their populations usually do not have sufficient funds to meet their basic needs with respect to water supply, sanitation and drainage.
- Lack of appropriate solid waste collection and disposal impacts on the water quality and the capacity of the urban drainage due to illicit garbage dumping. Desbordes and Servat (1988) stated that in some African countries there is no urban drainage, and where drainage systems do exist they are often filled with garbage and sediments. Tokun (1983) also mentioned this type of problem in Nigeria, where the drainage systems are used for garbage disposal.
- Lack of prevention programmes for occupation of risk-prone areas. When floods occur, funds are given to the local administration to cope with the problem, without any requirement to prevent its repetition in the future.
- Lack of knowledge for coping with floods at various levels of government administrations.
- Lack of municipal institutions dealing with urban drainage with respect to regulation, capacity building and administration (Ruiter, 1990).
- Dumping of solid waste into storm sewers; this may be encountered in developed countries as well.
- Unrealistic regulations for urban land use, in relation to social and economic conditions, may dissuade owners from complying with such regulations.

3.3.2 Stormwater

Stormwater is mostly rainwater that runs off impermeable surfaces in urban areas, including roofs, sidewalks, streets and parking lots. It is drained from urban areas by

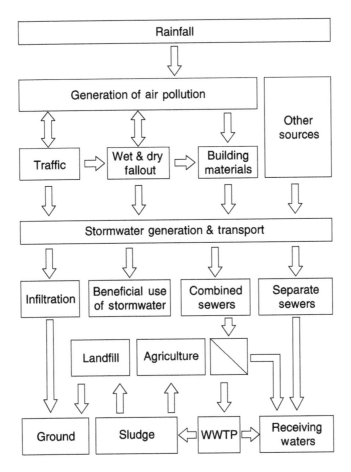

Figure 3.4 **Flows of water and pollutants in stormwater systems**

sewers or open channels to avoid local inundation. During this process, stormwater becomes polluted and its discharge into receiving waters causes environmental concerns. A schematic of runoff generation and pollution is shown in Figure 3.4.

Stormwater may be transported either by combined sewers, together with domestic and industrial wastewaters, or by separate sewers discharging to the nearest stream or lake. In combined sewers, high stormwater inflows can exceed the pipe capacity and excess flows have to be diverted by flow regulators as combined sewer overflows (CSOs) to the nearest receiving waters. CSOs contain not only the stormwater, but also untreated wastewater and sewer sludge; their direct discharge into receiving waters causes serious pollution problems.

The stormwater contribution to the wet weather flow reaching the wastewater treatment plant also increases the concentrations of heavy metals and other contaminants in the WWTP effluent and in the sludge (biosolids). This is one of the main reasons why the use of sludge from European WWTPs as a fertilizer in agriculture has been criticized.

Which sewer system is better – the separate system in which polluted stormwater may be discharged directly into the receiving water, or the combined system in which

Table 3.4 **Quality of urban runoff and combined sewer overflows**

Chemical constituent	Units	Urban stormwater		European CSO data (Marsalek et al., 1993)
		Mean of Duncan's dataset (1999)	US NURP Median site (US EPA, 1983)	
Total suspended solids (TSS)	mg/L	150	100	50–430
Total phosphorus	mg/L	0.35	0.33	2.2–10
Total nitrogen	mg/L	2.6	–	8–12
Chemical Oxygen Demand (COD)	mg/L	80	65	150–400
Biochemical Oxygen Demand (BOD)	mg/L	14	9	45–90
Oil and grease	mg/L	8.7	–	–
Total lead (Pb)	mg/L	0.140	0.144	0.01–0.10
Total zinc (Zn)	mg/L	0.240	0.160	0.06–0.40
Total copper (Cu)	mg/L	0.050	0.034	–
Faecal coliforms	FCU/100 mL	8,000	–	10^4–10^7

Sources: stormwater worldwide data, Duncan (1999); US NURP stormwater data, US EPA (1983); European CSO data, Marsalek et al. (1993)

the stormwater is conveyed to the WWTP for treatment, but CSOs occur? There are no general answers to this question, as real comparisons of separate and combined systems' performance depend on local conditions. The separate sewer system could be improved by developing separate stormwater treatment plants, but that may be costly and inefficient in view of highly variable infrequent inflows with low concentrations of pollutants. Similarly, the combined systems can be improved by incorporation of CSO pollution abatement measures.

3.3.2.1 Stormwater characterization

The literature on urban stormwater quality is very extensive. So far, more than 600 chemicals have been identified in stormwater and this list is growing. Makepeace et al. (1995) identified about 140 important contaminants which can be found in stormwater and would affect human health (mostly through contamination of the drinking water supply) and aquatic life. This list includes solids, trace metals, chloride, nutrients (N and P), dissolved oxygen, pesticides, polycyclic aromatic hydrocarbons and indicator bacteria. Typical concentrations of many such constituents have been reported in the literature, mostly for developed countries. Summaries of data from two large databases appear in Table 3.4.

Stormwater quality data have also been reported for less developed countries, but usually in small data sets. Examples of such data include those from Sao Paulo, Brazil (cited in Tucci, 2001), Johor, Malaysia (Yusop et al., 2004), Bandung, Indonesia (Notodarmojo et al., 2004) and Beijing, China (Che et al., 2004). In general, such data indicate significantly higher concentrations than those in Table 3.4, which may be caused by infrastructure problems, such as cross-connections between storm and sanitary sewers. In any case, such data suggest that pollution loads conveyed by storm sewers in developing countries are larger than indicated by the literature data published for developed countries.

Pollutant concentrations higher than the levels shown in Table 3.4 are observed during periods of snowmelt, when the pollutants accumulated in snowpacks are rapidly released and conveyed by storm sewers to the receiving waters (Viklander et al., 2003).

In older separate sewer systems, urban surface runoff is conveyed by storm sewers to the nearest receiving waters without any control or treatment. Only during the last thirty years, has stormwater management been introduced and practised by reducing runoff generation by allowing more rainwater to infiltrate into the ground, balancing runoff flows by storage and providing some form of runoff quality enhancement. Among the main pollutants of concern in stormwater, one could name suspended solids, nutrients (particularly P), heavy metals, hydrocarbons, and faecal bacteria.

3.3.2.2 Stormwater management

As a result of high discharges of stormwater, and the pollutant concentrations and loads that it conveys, and their potential impacts on the environment, alternative techniques – also referred to as best management practices (BMPs) – have been developed for stormwater management during the last several decades (Azzout et al., 1994; Baptista et al., 2005; Parkinson and Mark, 2005; Schueler, 1987; Urbonas, 1994), including the following:

- policies and source controls
- lot-level source controls
- biofiltration by grass filters and swales
- infiltration facilities
- bioretention
- green roofs
- water quality inlets and oil and sediment separators
- filters
- constructed wetlands
- stormwater ponds
- extended detention (dry) basins
- real-time control operation systems.

Policies and source controls are non-structural measures that are generally highly cost effective and as such are considered in all stormwater management plans. Typical examples of such measures are summarized below.

Public awareness, education and empowerment are essential for planning, design and acceptance of new stormwater facilities. Awareness and education are implemented through public meetings, open houses, tours of facilities and visual displays at stormwater management sites. In this process, concerned citizens/environmentalists groups are formed which then actively engage in environmental projects, including organized clean-up and publicity campaigns, and helping involve schools.

Urban development resource planning attempts to prevent problems resulting from urbanization. In this approach, new planning variables include population density and minimization of runoff from new developments by minimizing imperviousness and the associated impacts on receiving waters. Typical measures include progressive zoning ordinances and buffers for streams and wetlands. Material use, exposure and disposal controls strive to minimize the opportunity for contact between rainfall/runoff and various chemicals. This is generally achieved through good housekeeping, including proper storage of chemicals that could pollute runoff (e.g. road salts). In recent decades, a new term has been coined for this approach to stormwater management planning – low-impact development.

Spill prevention is practised to minimize the risk of spills during outdoor handling and transport of chemicals. Besides instituting good practices for chemical handling, measures for spill containment (berms, enclosures, separators) need to be developed. Illegal dumping and illicit connections need to be prevented or eliminated through public education (e.g. the yellow fish signs on sewer inlet grates), ordinances and their enforcement with penalties. This group of measures includes the management of both liquid and solid waste, including yard trash. Illicit connections should be prevented by enforcement of ordinances, and other sources than runoff should be disconnected from sewers.

Finally, drainage system performance needs to be maintained by proper street and storm sewer maintenance, and by adherence to BMP. Examples of maintenance procedures include street sweeping, catch basin cleaning, road and bridge maintenance, and specific maintenance measures recommended for individual BMPs.

Lot-level source controls represent minor measures implemented on site, mostly in the form of source controls. Such measures include enhanced rooftop detention, flow restrictions at catch basins to enhance local storage/detention, reduced lot grading to slow down runoff flow and enhance infiltration, redirecting roof leader discharges to ponding areas or soakaway pits, and sump pumping of foundation drains (MOE, 2003).

Biofiltration by grass filters and swales serves to reduce runoff volume by infiltration and to enhance runoff quality by such processes as settling, filtration, adsorption and bio-uptake. Swales are commonly applied in the upstream reaches of drainage areas to control runoff flows and enhance runoff quality. Vegetated filter strips are feasible in low-density developments with small contributing areas with diffuse runoff, suitable soils (good sorption), and lower groundwater tables. Swales are shallow grassed channels functioning in a similar way as vegetated biofilters and are best suited for small areas with permeable soils and low groundwater tables (Schueler, 1987).

The most common infiltration facilities are infiltration basins, pits, trenches and perforated pipes (MOE, 2003; Schueler, 1987; Urbonas, 1994). In general, such facilities infiltrate stormwater into the underlying soils or underground gravel-filled vaults. Some of these measures have been used for a long time on a small scale in rural settlements; only during the last few decades they have been further developed and used in urban areas on a larger scale. Porous pavement represents another infiltration measure for reducing surface runoff (Urbonas, 1994) and its effective use has been reported, for example, in France (with subsurface gravel-filled storage; Azzout et al., 1994). Stormwater infiltration helps keep the groundwater table at a natural level, which promotes good conditions for vegetation and a good microclimate. The construction costs of drainage systems with infiltration facilities are also cheaper than those of conventional systems. Infiltration is also implemented on grass or other permeable surfaces, and in drainage swales and ditches. The use of this measure is steadily growing in many countries.

Bioretention facilities combine treatment features of grass filter strips and infiltration facilities. Stormwater is pre-treated by running across the filter strips toward a depressed bioretention area, where it infiltrates into the soil. Where the native soils do not have sufficient infiltration capacity, they are replaced by engineered soils designed to infiltrate and store stormwater. Some water is returned to the atmosphere through evapotranspiration.

Green roofs, which are gaining on popularity in developed countries in certain climates, are rooftops planted with vegetation. For this purpose, a layer of soil (typically 0.15–0.3 m thick) is provided to grow low vegetation and thereby provide roof runoff control by storage of rainwater in the soil and by rainwater evapotranspiration. Generally, green roofs are effective in rainwater runoff control with respect to both quantity and quality, and provide an additional benefit in the form of evaporative cooling.

Water quality inlets were originally developed as small three-chamber storage tanks installed at inlets to the sewer system. They provide some stormwater treatment by sedimentation and skimming of floatables (and oil), and are particularly well suited for parking lots, and commercial or industrial land. Original designs were susceptible to washout of deposited materials during severe storms (Schueler, 1987). Oil/grit separators function similarly as water quality inlets, but can also be installed in-line at locations further downstream from inlets. A number of designs are available on the market and indicate good potential for removing coarse solids (sand) and containing free oil spills (MOE, 2003). On the other hand, the efficiency of these units in trapping fine sediments and chemicals attached to the sediments may be poor, because of under-sized units or a lack of flow-limiting devices that prevent the washout of trapped materials.

Stormwater (sand) filters are effective in removing pollutants, but attention must be paid to clogging (e.g. by some stormwater pretreatment) and backwashing may be required (Schueler, 1987). Simpler maintenance is achieved by breaking up (i.e. by raking) the surface layer, which may get clogged by formation of a biofilm. Good designs are equipped with a collector of the treated effluent and an overflow/bypass structure (Urbonas, 1994). Biofilters (with a coarse medium with biofilm grown on granular surfaces) were also tested and show good promise for removal of dissolved heavy metals (Anderson et al., 1997).

Constructed wetlands provide stormwater detention and treatment by various processes, including filtration, infiltration and biosorption, and remove both particulate and dissolved pollutants (Rochfort et al., 1997). Constructed wetlands are designed with shallow depths (0.15–0.6 m deep) and typically occupy less than 2% of the catchment area. Some problems associated with this BMP include thermal enhancement, seasonal variations in performance, poor performance in cold climates during winter months, and complicated maintenance (MOE, 2003).

Stormwater ponds have been used in many countries to attenuate drainage flows and treat stormwater by removal of suspended solids, heavy metals and, to some extent, nutrients. The cost of construction and operation of such facilities is often low compared with their environmental and community benefits. The sediments from ponds may contain high concentrations of heavy metals. In general, stormwater ponds should be recognized as treatment facilities and not as natural water bodies, even if they often provide aesthetic value to the urban area.

Extended detention (dry) basins provide stormwater settling in areas where it is difficult to maintain wet facilities. They are widely applicable. Dry ponds with deposited sediment may appear unattractive, and there may be concerns about exposure to contaminated sediments (Schueler, 1987).

Real-time control operation of sewer systems has been developed during the last two decades and implemented in some Canadian, European, Japanese and US cities. The applications are often in combined sewer systems and the purpose is mainly to reduce combined sewer overflows and/or overloading of wastewater treatment plants, by the maximum utilization of the dynamic capacities of the system (Colas et al., 2004).

3.3.2.3 Special considerations for drainage in cold climates

In countries with a cold climate (i.e. those with freezing temperatures over periods of several months) the precipitation falls as snow during a significant part of the year. When the snow is cleared from streets in the cities, it is either brought to local snow dump sites or to a central deposit site outside the city, or it is dumped into watercourses. When the snow melts, the meltwater runs off in the same way as stormwater. However, the impacts may be more severe due to the following facts:

- Snowmelt may generate high flows, causing surcharging of the sewer systems and possibly flooding in the receiving waters.
- The meltwater often has higher concentrations of heavy metals, sand and salt than stormwater (sand and salts being used for de-icing urban roads and streets).
- The impacts of urban snowmelt on streams, lakes and ponds may be exacerbated by ice covers of such water bodies and densimetric stratification. High salt concentrations and oxygen depletion have been noted at a number of locations (Marsalek et al., 2003).

Various technologies have been adopted for treating urban meltwater. Examples are ponds, oil and grit separators, and infiltration facilities. They all have to be designed and operated with consideration of the special effects of temperature, ice and snow conditions, and the higher pollutant concentrations (Viklander et al., 2003).

3.3.3 Combined sewer overflows

Even though combined sewer overflows (CSOs) are highly polluted, they are often discharged into nearby receiving waters without much treatment and cause serious pollution. The magnitude of annual CSO discharges depends on the extent of combined sewers (percentage of the total), climate, and design policies and practice. The extent of combined sewers varies from country to country, ranging from 20 to 90%. Generally combined sewers are more common in climates with lower annual rainfall; for high rainfalls, the system would be too overloaded and would collect only a low percentage of total flows. Finally, the overflow setting, typically in multiples of dry weather flow, greatly influences the spilled CSO volume. Typical settings vary from two to six times dry weather flow (Marsalek et al., 1993).

3.3.3.1 CSO characterization

The pollution characteristics of CSOs, while somewhat similar to those of stormwater, are strongly affected by domestic sewage and sewer sludge washout from combined sewers. Consequently, CSOs are particularly significant sources of solids, biodegradable organic matter, nutrients, faecal bacteria, and possibly some other chemicals originating from local municipal/industrial sources. During the early phase of runoff, referred to as the first flush, the CSOs' characteristics approximate to or even exceed pollutant concentrations in raw sanitary sewage. After the first flush, pollutant concentrations in CSOs subside. Their impacts on receiving waters are similar to those described in the preceding section, but much stronger in terms of oxygen depletion, eutrophication and increased productivity, and faecal pollution. It is desirable, therefore, to control CSOs prior to their discharge into the receiving waters.

CSOs are not routinely monitored, except for special studies of local significance. This follows from the diffuse and intermittent nature of these sources, for which large-scale monitoring programmes would be prohibitively expensive. Nevertheless, over the years, a fair number of studies attempting to assess these sources have been undertaken in a number of municipalities or regions. A summary of such data from European sources was presented earlier in Table 3.4.

In comparison to stormwater, the pollution strength of CSOs is similar to that of stormwater for total suspended solids (TSS), but greater for biochemical oxygen demand (BOD), indicator bacteria, total nitrogen (TN) and total phosphorus (TP), and generally smaller for unconventional pollutants, including heavy metals, polycyclic aromatic hydrocarbons (PAHs) and organochlorine pesticides.

3.3.3.2 CSO control and treatment

CSOs are caused by excessive inflows of stormwater into the sewer system, so any measure discussed in the preceding section for reducing stormwater runoff and its inflow into combined sewers would also help abate CSOs. Such helpful measures include all lot-level measures, infiltration measures (pits, trenches, basins, porous structures) and porous pavements (Urbonas, 1994). The mitigation of actual overflows is accomplished by various forms of flow storage and treatment; flow storage serves to balance CSO discharges, which may be returned to the treatment plant after the storm, when flows have subsided below the plant capacity (Marsalek et al., 1993).

CSO storage can be created in a number of ways: by maximizing the utilization of storage available in the existing system (e.g. through centrally controlled operation of dynamic flow regulators in real time – Schilling, 1989), as newly constructed storage on-line or off-line (on-line storage includes oversized pipes or tanks; off-line storage includes underground storage tanks or storage and conveyance tunnels), or even in the receiving waters (the flow balancing systems created by suspending plastic curtains from floating pontoons, in a protected embayment in the receiving waters) (WPCF, 1989). Some storage facilities are designed for treatment, more or less by sedimentation, which can be further enhanced by installing inclined plates. Stored flows are returned to the wastewater treatment plant, which must be redesigned/ upgraded for these increased volumes. Without such an upgrade, the plant may become overloaded and its treatment effectiveness impaired, and the benefits of CSO storage would be defeated.

CSO storage tanks have become a common design feature in many European sewerage systems, and are increasingly being used in North America. In this approach, an additional 'storage' volume for flow and pollutant load retention is included in the form of an oversized sewer pipe or storage chamber, which is incorporated into the sewerage at the points of overflow. Tanks are normally either on-line (continually in operation) or off-line (to which flow is diverted during high-flow periods via a diversion structure).

CSO treatment takes place either at the central plant, together with municipal sewage, or can be done in satellite plants dedicated to this purpose. Various processes have been proposed or implemented for the treatment of CSOs, including settling (plain, inclined plate and chemically aided), hydrodynamic separation, screening, filtration, dissolved air flotation, and the Actiflo™ process with coagulation and ballasted settling (Zukovs and Marsalek, 2004). Furthermore, the treated effluents may be disinfected, either by conventional chlorination (sometimes followed by dechlorination), or by UV irradiation (WPCF, 1989).

Treatment technologies are available to achieve almost any level of CSO treatment, but proper cost–benefit considerations are crucial for achieving the optimal level of CSO pollution abatement, within the given fiscal constraints. Reductions in treatment capacities are obtained by flow balancing of inflows by storage (Marsalek et al., 1993). From the maintenance point of view, the operators (municipalities) prefer relatively simple treatment systems, with more or less automatic operation, and minimum maintenance requirements.

The most cost-effective CSO abatement schemes deal with the entire urban area (and all system components) and represent combinations of various source controls, storage and treatment measures, allowing various degrees of control and treatment, depending on the event frequency of occurrence (Marsalek et al., 1993). More frequent events should be fully contained and treated; less frequent events may be still fully or partly contained and treated to a lower degree, and finally, infrequent events would still cause overflows, but of reduced volumes and could receive some pre-treatment prior to their discharge into the receiving waters.

The complexities of combined sewer systems, and the dynamics of flow, storage, loads and treatment processes, make it particularly desirable to control the sewerage/treatment/receiving water systems in real time. Real-time control (RTC) has been found particularly useful in systems with operation problems that vary in type, space and time, and that have some idle capacity (Schilling, 1989). The best-developed types of RTC are those for wastewater quantity and the associated modelling. The remaining challenges include RTC of quality of wastewater and receiving waters, and reliable hardware (Colas et al., 2004). It was suggested that in typical wastewater systems with no control (i.e. control by gravity only), approximately 50% of the system capacity remained unused during wet weather. By applying RTC, about half of this potential could be realized (Schilling, 1989).

3.4 WASTEWATER AND SANITATION

Urban populations require access to adequate sanitation and disposal of generated solid and liquid wastes. Such objectives are achieved by wastewater management and sanitation.

3.4.1 Problem definition

According to the *Global Water Supply and Sanitation Assessment 2000 Report* (WHO and UNICEF, 2000) the proportion of the world's population with access to excreta disposal facilities increased from 55% in 1990 to 60% in 2000. However, at the beginning of 2000, 2.4 billion people lacked access to improved sanitation. The majority of these people lived in Asia and Africa; for example, less than half of the Asian population had access to improved sanitation. Thus, a large part of the world population is still exposed to waterborne diseases and degraded water resources (Ujang and Henze, 2006). In developing countries, the percentage of the population served by wastewater treatment hardly reaches 15% (US EPA, 1992) and such treatment is almost always only a primary or inefficient secondary level treatment. However, there are differences among the countries in this group, with one category representing countries in transition with higher growth rates and rapid industrialization and urbanization, and the other category representing countries with slower growth rates. It is therefore important to adopt sanitation approaches that fit the local conditions and are site-specific.

Effluents are frequently used for agricultural irrigation or discharged into soils, rivers or lakes, and ultimately reach the sea. Agricultural reuse of treated wastewater effluents is the most common option in arid and semi-arid areas, because of the lack of water and also because farmers know the value of sewage for increasing soil productivity by input of nitrogen, phosphorus and organic matter. Nevertheless, this practice also considerably increases health risks.

The global sanitation coverage with 'improved sanitation' is estimated to be around 60%. The coverage in urban areas is significantly higher than in rural ones, ranging from 78% in Asia to almost 100% in developed countries. These numbers must be considered unreliable, due to the vague definition of urban areas and to the fact that the term 'improved sanitation' includes not only connections to public sewers but also to septic systems, pour-flush latrines, simple pit latrines, and ventilated, improved pit latrines.

Water quality in receiving waters is affected mainly by wastewater discharges, solid wastes and wet-weather flow pollution. The impacts of such discharges depend on the type of pollutant, the relative magnitude of the discharge compared with the receiving waters, and the self-purification capacity of receiving waters.

In developed countries, even though wastewater is treated at a secondary level, treated effluents still pollute.

The challenge now is to provide the world's population, especially the poor, with adequate water and sanitation facilities. The future tendencies indicate further deterioration of water resources, because of a growing world population and continuing environmental degradation due to poor pollution control, particularly in most of the developing countries. In order to advance water and wastewater management in developing countries, it is necessary to consider the segments of the society itself, particularly the types of housing areas, and segregate the funding framework of individual proposed projects. Higher-income urban communities are generally willing to pay for sewerage services and appropriate sanitation systems can be provided.

During the past decade, there has been much debate about the 'conventional sanitation' approach and its applicability in developing countries. Consequently, new sanitation concepts have been introduced under such headings as ecological sanitation (ecosan), or ecologically sustainable sanitation, with both terms used interchangeably. The use of sustainable sanitation is essential for achieving the Millennium Development Goals by 2015 with respect to providing water supply and adequate sanitation for developing countries. Furthermore, sustainable sanitation is flexible enough to allow for differing applications in any community, regardless of its wealth or water availability, and requires lower investment costs than conventional sanitation. Finally, sustainable sanitation is much easier to adopt in developing countries where water supply and sanitation infrastructures are still in the developing stages, and an infrastructure concept can be adopted (Ujang and Henze, 2006).

3.4.2 Technological development

Some milestones in the development of modern, centralized wastewater systems are summarized below (Wolfe, 2000).

Around 3500 BC, brick and stone stormwater drain systems were constructed in streets of Mesopotamia. In Babylon access-holes and clay piping were used to connect in-house bathrooms to street sewers around 3000 BC. In Rome the famous *cloacas* were constructed, starting in 600 BC (*Cloaca Maxima*). The construction of sewers in Paris started in the fourteenth century. An underground sewer for draining cellars and carrying away wastewater was constructed in Boston around 1700. The city of Hamburg is credited with having built the first city sewerage system around 1840. Sewage pumps were introduced around 1880, followed by the development of screens and grit chambers to remove solid matter. The modern WC (water closet) was developed by Thomas Twyford in 1885.

Wastewater treatment technologies were successively developed from the end of the nineteenth century. Among the important developments were contact beds (1890), trickling filters (1901), sludge digestion (1912), the activated sludge process (1913), and surface aeration (1920). Biological as well as chemical treatment systems have been steadily improved and installed, removing solids, organic matter and nutrients from the wastewater. A schematic of such a conventional system is shown in Figure 3.5. At the beginning of the twenty-first century much interest is being paid to combinations of biological treatment and different kinds of microfiltration (membrane technology). The protection of the environment from harmful discharges of wastewater is governed by laws and directives, for example, the European Union Waste Water Treatment

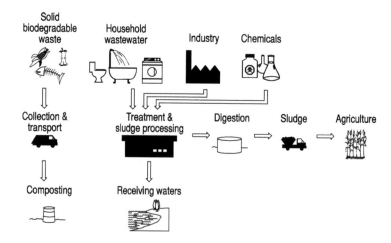

Figure 3.5 **A conventional system for managing wastewater and organic waste**

Directive 91/271/EEC (Council of European Communities, 1991), and many national regulations.

3.4.3 Ecological sanitation

There is a fundamental connection between the present state of water supplies, sanitation, solid (organic) waste management and agricultural development worldwide. While sustainable provision of water and sanitation for a growing population is in itself a formidable challenge, the new target is to develop technologies and management strategies that can make organic residuals from human settlements useful in rural and urban agriculture for production of food. Traditional methods used in water resources development and in provision of sanitation were, and still are, unable to satisfy the fast-growing needs of developing countries.

In other words, close coordination between urban and rural components for nutrient and biomass recycling is required, particularly for a sustainable society. After consumption, the question is how to deal with urine and faeces and utilize them in the recycling of nutrients and biomass. One solution is a separation of urine from other household wastewater. This process is called ecological sanitation – a dry toilet with separation of urine and faeces at the source (Winblad and Simpson-Hebert, 1998). The separated urine can then be transferred to the nutrient processing plants, where nitrogen and phosphorus are recovered and transformed into chemical fertilizers.

If the urine and faeces are collected and used in agriculture, the remaining wastewater, known as greywater, is easier to treat by conventional wastewater treatment processes. Pipe systems are not necessary for collecting greywater and open channels are still acceptable for minimum conditions. The collected greywater can be treated locally and discharged to surface waters in the vicinity of the collection points (Maksimovic and Tejada-Guibert, 2001). The nutrient content of greywater is comparable to waters that by different standards are regarded as 'clean' (Gunther, 2000). However, there are concerns about pharmaceuticals and personal care products and therapeutics in greywater, and these need to be considered in greywater reclamation and reuse (Ternes and Joss, 2006). Also sludge revalorization causes some concerns due to associated health risks, particularly in developing countries. The spread of diarrheic diseases can be a health concern unless the pathogen content in the ecosanitation sludge is greatly reduced (Jiménez et al., 2007). That is why EcoSanRes (2005) and Schönning and Stenström (2005) recommend that, even if treated, ecosanitation black matter (faeces) should be handled safely and not used to fertilize vegetables, fruit or root crops that will be consumed raw.

A greywater reuse system should be able to receive the effluent from one or more households during all seasons of the year. Where garden soils of low permeability become saturated by winter rainfall, there should be opportunities to divert excess water to sewers or to arrange an alternative form of disposal. The reuse system needs to protect public health, protect the environment, meet community aspirations and be cost effective (Anda et al., 1996).

The environmental, economic and public health advantages of ecosan are indisputable, and the basic technologies are well known. For broad implementation, it is necessary to provide and disseminate the respective educational tools, as demonstrated in the UNESCO and GTZ report on capacity building for ecological sanitation (UNESCO and GTZ, 2006).

Esrey et al. (2001) summarized the basic characteristics of ecological sanitation systems as follows:

- *Preventing disease*: must be capable of destroying or isolating faecal pathogens.
- *Protecting the environment*: must prevent pollution and conserve valuable water resources.
- *Returning nutrients*: must return plant nutrients to the soil.
- *Culturally acceptable*: must be aesthetically inoffensive and consistent with cultural and social values.
- *Reliable*: must be easy to construct and robust enough to be easily maintained in a local context.
- *Convenient*: must meet the needs of all household members whatever their gender, age and social status.
- *Affordable*: must be financially accessible to all households in the community.

3.4.4 Basic demands on wastewater management systems

The basic needs to be met by an urban wastewater management system are pollution control, public health protection, avoidance of flooding and recycling of nutrients.

Pollution control aims to protect the receiving waters, including creeks, streams, rivers, lakes and the sea, against discharges of wastewater that may cause eutrophication, oxygen depletion, toxicity and other negative impacts that decrease biological diversity or impair beneficial uses of the receiving waters for such purposes as drinking water supplies for downstream settlements. Pollution control is one of the major reasons for the construction of WWTPs.

Public health protection has historically been the major driving force behind the construction of WWTPs. In the West as well as in developing countries, the application of wastewater treatment has resulted in drastic improvements in the health standards of people living in affected areas. In particular, the introduction of disinfection has been successful, as further elaborated below.

Flooding in urban areas, which is caused either by locally generated floods or by floods generated in the upstream catchment, was discussed earlier in Section 3.3.1. Local (pluvial) flooding can be exacerbated by poorly designed and functioning sewer systems. Flood damage caused by local flooding is often less severe in drainage systems that use open channels, ditches or swales for flow conveyance, rather than underground sewers.

Recycling of nutrients has been included in the basic needs since the sustainability requirements were formulated at the United Nations-sponsored meetings in Oslo, Rio and Johannesburg. The nutrient contents of domestic wastewater may be valuable as fertilizer in agriculture (or aquaculture). Associated problems are the contamination of the sludge with heavy metals and certain trace organic substances.

Additional demands on wastewater systems are that they should be affordable, acceptable to the public and convenient. Affordable wastewater systems provide services which the users are able to pay for. The distribution of service costs is essential, and the rates should be equitable and fair to all connected. The consumers must be willing to pay the costs. As a general guideline, the World Bank recommends that the cost of water and sanitation should not exceed 5% of total family income.

Table 3.5 **Typical composition of sewage**

Parameter	Concentration		
	Minimum	*Average*	*Maximum*
Total solids (mg/L)	350	720	1,200
Dissolved solids, total (mg/L)	250	500	850
Fixed solids (mg/L)	145	300	525
Volatile solids (mg/L)	105	200	325
Total suspended solids (mg/L)	100	220	350
Fixed solids (mg/L)	20	55	75
Volatile solids (mg/L)	80	165	275
Settleable solids (mL/L)	5	10	20
Biochemical Oxygen Demand (BOD_5) (mg/L)	110	220	400
Chemical Oxygen Demand (COD) (mg/L)	250	500	1,000
Total Organic Carbon (mg C/L)	80	160	290
Total nitrogen (mg N/L)	20	40	85
Organic nitrogen (mg N/L)	8	15	35
Free ammonia (mg N/L)	12	25	50
Total phosphorus (mg/L)	4	8	15
Grease and oil (mg/L)	20	100	150
Alkalinity (mg $CaCO_3$/L)	510	100	200

Source: after Metcalf and Eddy, 2003.

The services provided must also be acceptable to the consumers, in terms of the service delivery, water quality and prices. The water and wastewater systems must be socially and culturally acceptable to the consumers. Different traditions exist in each country or region. For example, dry sanitation is not acceptable in some cultures.

The systems must also be convenient to use. The carrying of drinking water or wastewater products has strong gender implications and should be avoided in sustainable systems.

3.4.5 Wastewater characterization

Municipal sewage is a mixture of domestic, commercial and industrial wastewaters. In developed countries, industrial wastewaters are pre-treated prior to discharge into municipal sewers; in developing countries, industrial wastes are often not treated at all. In spite of this, the impact of industrial discharges in developing countries is smaller than in developed ones, because of a lower level of industrialization. The typical composition of municipal sewage is described in Table 3.5.

As shown in Table 3.6, pathogen concentrations in developing countries' wastewater are much higher than those in wastewater in developed countries.

The characteristics of industrial wastewaters vary greatly, depending on the type of industry. To protect receiving waters from industrial pollutants, it is important to develop efficient pre-treatment programmes which reduce or eliminate pollutants from industrial wastewater prior to discharge into municipal sewage systems. Without such pre-treatment, industrial pollutants might interfere with the operation of conventional wastewater treatment plants, or pass through them without much abatement and cause

Table 3.6 **Pathogen concentrations in wastewater in developing and developed countries**

Micro-organism	Developed countries	Developing countries
Salmonella (MPN/100 mL)	10^3–10^4	10^6–10^9
Enteric viruses (PFU/100 mL)	10^2–10^4	10^4–10^6
Helminth ova (HO/L)	1–9	6–800
Protozoa cysts (organisms/L)	28	10^3

Source: Chávez et al., 2002.

damage in receiving waters. Without such pre-treatment, industrial pollutants might pass through conventional WWTPs without much abatement and cause damage in receiving waters.

3.4.6 Wastewater systems without separation of wastewaters at the source

Systems that do not separate wastewaters at the source manage the total mixture of wastewaters, including blackwater and greywater. In terms of system architecture, they can be designed as conventional centralized systems, or less common distributed systems.

3.4.6.1 Centralized systems

Currently, most cities in all parts of the world have a centralized sewerage system that provide some kind of treatment. Exceptions are the poorest cities and unofficial 'squatter' settlements in peri-urban areas in Africa, Asia and Latin America. They all have the same basic features: collection of the wastewater in or near the houses, transport by gravity sewers or pressure sewers to a treatment plant, and discharges to the receiving waters by longer or shorter outfalls. In many of the old systems, domestic and industrial wastewater is mixed with stormwater in combined sewers. In principle, wastewater collection and treatment systems are similar in all large cities of the world. The degree of use of modern technologies may vary, however, depending on financial resources and political will.

It is reported from many developing countries that although they have access to central sewer systems and sewage treatment, the facilities do not operate properly and in some places not at all due to the lack of management, maintenance, funding and training. Furthermore, the costs of implementing a centralized system in mega-cities are exorbitant.

Currently, the research and technology advances that are of interest involve:

- further development and refinement of biological methods for nutrient removal from wastewater and the recovery of nutrients
- the use of membrane technology for wastewater treatment
- development of anaerobic methods for sludge digestion and treatment
- safe disposal/application of biosolids on agricultural land
- incineration of sewage sludge
- control of new chemicals of concern, including endocrine disruptors, pharmaceuticals (including antibiotics), and personal care and therapeutic products.

3.4.6.2 Distributed (local) systems

Conventional (centralized) sewerage systems are common in the central parts of most large cities, but the approaches to wastewater management are quite different in small towns or in the peri-urban areas on the outskirts of cities. Open canals are commonly used to transport human wastes and discharge them to receiving waters without treatment. Such systems are a severe threat to public health since the water in receiving streams, rivers or canals may be used further downstream for cleaning and washing, or even as a drinking water source. In densely populated areas, polluted rivers cause the same kind of a threat.

Recent research into centralized and decentralized systems indicates that the latter may be more sustainable, particularly when examining their applications in remote and rural areas, with integration of other technology sectors (energy, food production). Similar opportunities exist in new urban developments where there is high potential for using on-site, small and community-scale technology, which should be modified to reduce costs, but be based on the same science and research as the successful applications in developed countries (Ho, 2004, 2005).

Because of difficulties with recycling water and nutrients (i.e. closing their cycles), the current large-scale systems for supplying water, and collecting and treating wastewater, are not environmentally sustainable. On the other hand, small-scale water and wastewater systems can provide effective solutions in developing countries and achieve sustainability in both developed and developing countries. The technological sophistication of such systems is under discussion, with some authors (Wilderer, 2004) advocating the development and application of high-tech on-site treatment plants. With mass production, the costs of such package plants should be relatively low. In order to keep the plants operating properly, they should be controlled by remote sensing, and maintained by specialized service enterprises. The plant effluent should be hygienically safe and could subsequently be used for flushing toilets, washing clothes, cleaning floors or watering lawns.

In some parts of the world, simpler and cheaper solutions are sought for the management of wastewater on a local scale. Examples of such technologies are listed below:

- *Wastewater infiltration.* After sedimentation, the wastewater is infiltrated into the soil in a constructed filter plant, seeping down to the groundwater. The reduction of organic matter, nutrients and bacteria may not reach the standards of a high-tech WWTP, but may be a significant improvement on the existing conditions. This technology should not be used for industrial effluents or for wastewaters with high contents of dissolved substances (heavy metals, organic toxic compounds, pharmaceutical residues, etc.). Also, wastewater infiltration should not be used where the affected groundwater is used as a source of drinking water.
- *Constructed wetlands.* The term 'wetlands' is commonly used for many technologies: simple open ponds, several ponds in a series with or without vegetation, reed-beds with horizontal or vertical flow, and others. In the developed countries, wetlands are most commonly used as a polishing step after a WWTP, mainly for nutrient removal. The micro-organisms in the wetland nitrify and denitrify nitrogen to a certain degree, depending on the size of the wetland, the type of vegetation and the ambient temperature. In the Nordic countries, the use of wetlands is

increasing as a low-cost, natural technology, but the efficiency of the treatment is low during the winter. The major drawback of this technology is that it requires a large space, but this is not a universal limitation (Jayakumar and Dandigi, 2003).

• *Treatment ponds.* These have been used for a very long time and their performance has been improved considerably. Further improvements of their performance, especially in cold climates, have been achieved and demonstrated by addition of precipitation chemicals. Good guidance for pond treatment technology can be found in Shilton (2005).

• *The Living Machine.* In a few places in North America, domestic wastewater is transported to hydroponics plant beds in a greenhouse and subjected to treatment, while at the same time the nutrients in the wastewater are used for cultivating various green plants. Experimental facilities based on a similar concept also operate in Europe (Todd and Todd, 1993).

3.4.7 Systems with separation of wastewaters at the source

Originating in Sweden in the 1990s, the idea of separating domestic and other wastewaters at the source was developed and tested on a small scale. Since then, this concept has spread and facilities exist in many other European countries, including Denmark, Germany, Switzerland and the Netherlands, although in all cases only on a small scale. Several ongoing research projects are working on the refinement of these systems. Separation systems have also been applied in many places in developing countries, especially in rural areas where the alternative is pit latrines or no sanitation at all.

The advantages of ecological sanitation in rural areas in developing countries are striking: hygienic conditions are improved compared with simpler solutions, water is only used for cleaning purposes, and the wastewater products, urine and/or excreta, can be used as fertilizers after some minimum period of storage. The characteristics of various domestic wastewater sources are listed in Table 3.7. Although the chemical characteristics of various household wastes in developed and developing countries are similar, the biological characteristics differ greatly. For example, in developing countries, the helminth ova content in untreated faeces may be as high as 3,000 ova/g total

Table 3.7 Composition of urine, faeces, greywater, household wastewater and compostable household waste in Sweden (in kg/person equivalent/year)

Parameter	Urine	Faeces, incl. toilet paper	Greywater total	Household wastewater	Compostable household waste
TSS	7	19	26	53	25
VSS	3	17	15	35	21
COD_{tot}	3	23	23	49	34
BOD_7	2	12	12	27	12
N_{tot}	4.0	0.5	0.6	5.1	0.6
P_{tot}	0.33	0.18	0.25	0.76	0.10
S_{tot}	0.26	0.06	0.17	0.48	0.05
K_{tot}	0.88	0.33	0.29	1.49	0.23

Source: Jönsson et al., 2005.
Note: TSS, total suspended solids; VSS, volatile suspended solids; COD, chemical oxygen demand; BOD, biochemical oxygen demand.

solids (Strauss et al., 2003) and in treated ones from 0 to almost 600 ova/g total solids (Jiménez et al., 2006). A similar point was also emphasized in Section 3.4.5.

The use of separation systems in densely populated areas has been discussed but not tested in reality. Opponents argue that the cost of redesigning the sewerage systems in houses and in the streets will be huge, and that the transport of collected urine and/or excreta will cause additional costs, nuisance and air pollution in the cities. Advocates argue on the other hand that the cost–benefit balance of the system is favourable when compared with conventional systems, particularly after accounting for the recycling of nutrients and the use of natural resources. It would seem that the separation system may be feasible in peri-urban areas and 'informal' settlements where sewerage systems still do not exist and the extension of the central system to the outlying parts of the city would be too costly or almost impossible due to the lack of space and financial resources. A schematic of blackwater separation is shown in Figure 3.6.

3.4.8 Wastewater treatment technologies for developing countries

Modern wastewater management is based on cleaner production principles, including three intervention steps:

1. Minimize wastewater generation by drastically reducing water consumption and waste generation.
2. Provide treatment and optimal reuse of nutrients and water at the lowest possible level, such as on site and at the community level.
3. Enhance the self-purification capacity of receiving water bodies (lakes, rivers, etc.) through intervention.

The success of this approach requires systematic implementation, providing specific solutions to specific situations, based on appropriate planning, legal and institutional responses (Nhapi et al., 2006).

In developing countries, the kind of energy-intensive electromechanical wastewater treatment that is common in industrialized countries is too expensive and too difficult to operate and maintain. Consequently, more appropriate technologies should be used, including waste stabilization ponds, upflow anaerobic sludge blanket reactors, and waste-water storage and treatment reservoirs (or some combination of these). High-quality effluents from these processes can be reused for crop irrigation and fish aquaculture.

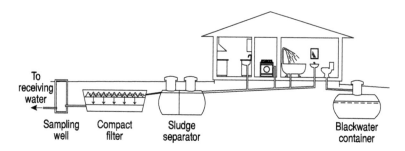

Figure 3.6 **Blackwater separation**

The warm climate in most developing countries makes the use of anaerobic technology applicable in domestic sewage treatment and less expensive. Compact systems employing sequential batch reactors (SBR) or dissolved-air flotation (DAF) systems are applied for the post-treatment of anaerobic reactor effluents and can achieve high performance in removing organic matter and nutrients from domestic sewage at ambient temperatures (Foresti, 2001).

For implementing municipal wastewater treatment in China, four technologies were proposed and found promising: natural purification systems, highly efficient anaerobic processes, advanced biofilm reactors, and membrane bioreactors (Qian, 2000a, 2000b). Other promising processes include the upflow anaerobic sludge blanket (UASB) or the anaerobic baffled reactor (ABR) as an anaerobic pre-treatment system, and the reed bed or stabilization pond with supporting media as a post-treatment system. Results obtained in pilot and full-scale treatment plants indicate that the anaerobic treatment is an attractive option for municipal wastewater pre-treatment at temperatures exceeding 20°C in tropical and subtropical regions. Both the UASB and the ABR systems have been employed as anaerobic pre-treatment systems. The effluents from the anaerobic treatment system should be post-treated, for instance by a reed bed system or in a stabilization pond packed with attached-growth media (Yu et al., 1997).

3.4.9 Case study of water pollution control in the Tehran Metropolitan Area

Water resources in the Tehran region include water stored in three reservoirs and the Tehran Aquifer, as well as local rivers and channels which are mainly supplied by urban runoff and wastewater. The sewer system employs the traditional soil treatment wells, and consequently the return flow from domestic use (more than 60%) is one of the main sources of groundwater recharge and pollution. Some portion of sewage is drained into local rivers and drainage channels and partially contaminates surface runoff and local streamflow. These polluted surface waters are used in conjunction with groundwater for irrigation purposes in the southern part of the region. Various decision makers and stakeholders, with conflicting interests, are involved in water pollution control in the study area (Karamouz et al., 2005a). In this case study, the Nash bargaining theory (NBT) was used to resolve the existing conflicts and provide the surface and groundwater pollution control policies, considering different scenarios for the development projects (Karamouz et al., 2005b). The results demonstrate the importance of an integrated approach for controlling pollution of surface waters and groundwater resources in the Tehran region.

Tehran annual domestic water use by more than eight million inhabitants is close to a billion cubic metres (Karamouz et al., 2004). The groundwater in the Tehran Aquifer, which is used as a source of drinking water, is polluted by the return wastewater flow from domestic soil treatment wells. In order to overcome the problems, several development plans are currently being investigated and implemented (Karamouz et al., 2005a). In a completed project, a network of more than 100 drainage wells was drilled to lower the groundwater table in the southern part of the city. The pumped well water is discharged to the local streams and channels and contributes to the flow of surface water in the southern part of the city. For more detailed information about surface water resources in the study area, see Karamouz et al. (2005a, 2005b).

Optimal operating policies for drainage wells as well as the optimal coverage of the Tehran Wastewater Collection Project (TWCP) have been developed for different scenarios in the development stages, taking account of the objectives and utility functions of the decision makers and the stakeholders of the system.

A conflict resolution model has been established (Karamouz et al., 2005b), taking account of the objectives of the following four stakeholders:

- Department of the Environment, concerned about water quality in the Tehran Aquifer
- Tehran Water and Wastewater Company, concerned about domestic water supply and wastewater collection and disposal
- Tehran Health Department, concerned about the drinking water quality standards
- Tehran Water Supply Authority, concerned about meeting water demands.

The conflict resolution model presented in this study is used to determine the optimal area that should be covered by TWCP as well as the pumping discharge of drainage wells in the development stages of the project. The results of the proposed model for different planning horizons indicate that the population covered by TWCP will reach 7.4 million people in the year of 2006. By maintaining the discharge capacity of the drainage wells at 50 million cubic metres per year, the quality of groundwater and also the groundwater table variation can be effectively controlled during the planning horizon (Karamouz et al., 2004).

3.4.10 Water and wastewater reuse

Water scarcity in many parts of the world puts stress on water supplies when the demand for water exceeds the available amount during certain periods, or when poor quality restricts the use of water. Water stress causes deterioration of fresh water resources in terms of quantity (aquifer over-exploitation, dry rivers, etc.) and quality (eutrophication, organic matter pollution, saline intrusion, etc.) (EEA, 1999). In urban areas this has emphasized the need to develop other kinds of water resources, such as desalination of seawater, collection of rainwater and reclamation of used water.

The reuse of treated wastewater has been practised in many countries over a long time, mostly for recharge of over-exploited aquifers by infiltration. Untreated wastewater is also used in some regions of the world due to the shortage of water and lack of economic resources to treat wastewater before reuse. It is estimated that at least 21 million ha are irrigated with treated, diluted, partly-treated or untreated wastewater (Jiménez and Asano, 2004). In some urban areas, wastewater has been used for agricultural irrigation. 'Urban agriculture' is practised in urban and peri-urban areas of arid or humid tropic countries where wastewater is available and there is local demand for fresh produce and a need to support people living on the verge of poverty. Wastewater flowing in open channels is used to irrigate very small plots of land where trees, fodder or any other crops can be grown; these may be brought to the market in small quantities (flowers and vegetables) or be used as part of the family diet (Cockram and Feldman, 1996; Ensink et al., 2004a). A tenth or more of the world's population consumes crops irrigated with wastewater (Smit and Nasr, 1992). The use of wastewater can vary considerably from one region to another; for example, in

Hanoi, Vietnam, up to 80% of locally produced vegetables are irrigated with waste-water (Ensink et al., 2004b).

A recent review by Bixio et al. (2005) analysed over 3,300 water and wastewater reclamation projects and concluded that technological risks no longer represent a major concern in reclamation and reuse, with the current main concerns focusing on financing, failure management and social acceptance.

More recently, direct use of treated stormwater and wastewater has been introduced in some countries in most regions of the world. On a small scale, reuse of stormwater and greywater has been applied in some European countries (the Netherlands, United Kingdom and Denmark) for garden irrigation and for flushing of toilets. In some cases, certain drawbacks have been observed; most are due to cross-connections with the drinking water supply system, causing risks of spreading waterborne diseases. In 2003 the use of treated greywater was prohibited in the Netherlands for this particular reason.

In other parts of the world, the need to develop complementary water resources has forced water authorities to go one step further by applying high-tech treatment and delivering reclaimed water to customers for various uses. Some examples are given below.

Various types of water, including reclaimed wastewater, are used in aquaculture to produce fish and grow aquatic crops. Some guidance can be obtained from WHO guidelines (WHO, 1989), which are currently under review. A preview of the new draft WHO guidelines for water reuse in agriculture was published by Kamizoulis (2005) and noted that the new guidelines provide basic information on health risks and a methodology for developing health-based targets by quantifying the risk and developing pathogen reduction targets. Health risk control (protection) measures are also presented, and include wastewater treatment technologies, crop restrictions, wastewater application methods and human exposure controls.

3.4.10.1 NEWater in Singapore

In February 2003, Singapore started to replenish about 1% of its total daily water consumption with reclaimed wastewater, which was named NEWater. NEWater is mixed and blended with raw water in the reservoirs before undergoing conventional treatment at the waterworks for supply to the public for potable use. The amount will be increased progressively to about 2.5% of total daily water consumption by 2011. Singapore suffers from a water shortage and buys more than half of its water demand from neighbouring Malaysia under decades-old treaties, which will start expiring in 2011. The water trade has caused many arguments between the two nations over pricing and other issues.

The wastewater used for NEWater is the product of a multiple-barrier water reclamation process. Treatment steps are conventional wastewater treatment, microfiltration, reverse osmosis and finally ultraviolet disinfection. The quality of the reclaimed water fulfils all requirements and is in most aspects better than the raw source water currently used (more information can be obtained from the Public Utilities Board and the Ministry of the Environment and Water Resources, Singapore, www.pub.gov.sg/NEWater/).

3.4.10.2 Shinjuku water recycling centre, Tokyo, Japan

The water recycling centre in the Shinjuku district of Tokyo distributes advanced treated wastewater for toilet flushing in high-rise office buildings in the districts of

Shinjuku and Nakano-Sakaue. The reclaimed water is supplied from the Ochiai WWTP, located two kilometres from the water recycling centre. The treatment process applied in reclamation includes conventional secondary wastewater treatment (primary sedimentation, activated sludge process and secondary sedimentation) supplemented by rapid sand filtration. At present, the water recycling system operates in twenty-six high-rise buildings. Differentiated fees for water supply were implemented to support the choice of recycled water, rather than municipal drinking water, for toilet flushing (Asano et al., 1996).

3.4.10.3 Wetlands with fish production in Calcutta, India

The city of Calcutta, India, has ancient wetland traditions, and the existing techniques are used for direct recycling of nutrients in the city, which is one of the largest in the world. The sewage system in central Calcutta was constructed in the 1870s and is completely inadequate today. Some parts of the city built in later stages of the city development have no sewers at all. There is also frequent flooding during the monsoon period. Currently, there is one WWTP operating in Calcutta, and two more plants are under construction.

Most of the wastewater in Calcutta is untreated today, and will remain untreated when the ongoing project is completed. However, some of the untreated wastewater is diverted into Calcutta's wetlands, which are part of the largest aquaculture area in the world. Wastewater is pumped directly to huge wetlands where fish are cultivated and sold on the market. The use of wetlands has a centuries-old tradition in India. In the 1980s, systematic follow-up studies of the wetlands were initiated, under the auspices of the newly established Institute for Wetland Management and Ecological Design (IWMED). The area is enormous, corresponding to 12,000 ha, and produces a sixth of all the fish consumed in Calcutta (Ghosh, 1999).

3.4.10.4 Reuse of (untreated) sewage for agricultural irrigation in the Mezquital Valley (Mexico City sewage disposal)

Mexico is a country with apparent water sufficiency at the national level. However, two-thirds of the territory suffers from a lack of water. Frequently, municipal wastewater is used for irrigation. In 1995, a total of 102 m³/s of wastewater was used to irrigate about 257,000 ha in the country. An example of this practice is Mexico City sewage, which has been used to irrigate the Mezquital Valley, north of the city, since 1896; 52 m³/s of entirely untreated wastewater has been used to irrigate several crops and has made possible the economic development of the region. This is the largest and oldest scheme of agricultural irrigation using urban wastewater in the world (Mara and Cairncross, 1989). As a result of this practice, the water table of the aquifer underlying the irrigation zone has been rising. The unplanned artificial recharge is about 25 m³/s, and springs with capacities of 100 to 600 L/s started appearing thirty-five years ago (Jiménez and Chávez, 2004). This 'reclaimed water', treated only with chlorine, is being used to supply 300,000 inhabitants of the region for human consumption. Several studies have shown that the water meets potable norms and another 288 parameters from the WHO human consumption guidelines, including toxicological tests (Jiménez et al., 2001). Also, this wastewater is enriched in nutrients, as confirmed by increased yields of agricultural crops (see Table 3.8).

Table 3.8 Yield increase as a result of irrigation with wastewater in the Mezquital Valley, Mexico

Crop	Yield (tonnes/ha)		Increase (%)
	Wastewater	*Fresh Water*	
Maize corn	5.0	2.0	150
Barley	4.0	2.0	100
Tomato	35.0	18.0	94
Oats for forage	22.0	12.0	83
Alfalfa	120.0	70.0	71
Chili	12.0	7.0	70
Wheat	3.0	1.8	67

Source: Jiménez et al., 2005.

3.4.10.5 Reuse of stormwater and greywater in Sydney, Australia

Serious summertime water shortages in Sydney, Australia, prompted Sydney Water to encourage the reuse of stormwater and greywater, besides the ongoing intensive water saving and conservation campaigns. The use of reclaimed water for garden irrigation is encouraged, either by using treated greywater or stormwater collected in separate tanks. Advice to house owners can be found on their website, www.sydneywater.com.

Sydney Water also introduced the distribution of reclaimed water. One example is the Rouse Hill Development area. These homes have two water supply systems, reclaimed water and drinking water. To ensure that the drinking water is not confused with the recycled water it is delivered by a separate distribution system. The reclaimed water taps, pipe-work and plumbing fittings are coloured lilac for easy identification. Reclaimed water is passed through a complex treatment train including ozonation, microfiltration and chlorination, in addition to the customary high level of treatment. The reclaimed water is subject to strict guidelines that limit its use to toilet flushing and outdoor purposes, such as car washing and garden irrigation (NSW Health, 2000).

3.4.11 Closing observations on wastewater management in developing countries

Municipal wastewater management in developing countries focuses on protection of both public health and the environment (Ujang and Henze, 2006). In general, during the past twenty-five years, the development of wastewater management in the developing countries of Asia and the Pacific has generally lagged far behind the development of water supplies, partly due to the lack of funding and partly due to government policies. Efforts to improve hygiene by good water supplies are undermined if complementary sanitation and wastewater management are not also provided.

Historically, over a fifth of the Asian Development Bank's (ADB) overall lending has been in the water sector. From 2002 to 2005, ADB lending for water supply, sanitation and wastewater management totalled over $1.5 billion, with about $230 million spent on wastewater management. From 2006 to 2008, this figure will rise to more than

$2.3 billion, with about $600 million reserved for wastewater management (Lohani, 2005). A recent study estimated that it will cost $8 billion a year until 2015 to meet the regional Millennium Development Goals for water supply and sanitation, and double that amount to cover all the population without any services. That is a significant but not an impossible figure.

The high investments needed raise the question – how can one advance wastewater management without being too constrained by the huge investment required? The answer can be found in the Millennium Development Task Force definition of basic sanitation as

> the lowest-cost option for securing sustainable access to safe, hygienic, and convenient facilities and services for excreta and sullage disposal that provide privacy and dignity, while at the same time ensuring a clean and healthful living environment both at home and in the neighbourhood of users.

Thus, one needs to look for low-cost options that can be adopted while governments build their capacity to finance full sewerage. The specific technologies that meet local urban and rural conditions may vary from place to place: in dispersed, low-income rural areas, the appropriate technology may be a simple pit latrine, whereas in a congested urban slum area with a reliable water service, it may be a low-cost sewerage system (Lohani, 2005).

Chapter 4

Impacts of urbanization on the environment

Urbanization affects many resources and components of the environment in urban areas and beyond. Even though the discussion in this report focuses on water, brief discussions of some connected issues, or other environmental compartments interacting with urban water, cannot be avoided. Thus, for the sake of completeness, the discussion covers, in various degrees of detail, the impacts of urban areas on the atmosphere, surface waters, wetlands, soils, groundwater and biota.

Urban areas produce air pollution and release heat into the atmosphere. In turn, air pollution of local or remote origin may be a significant source of pollutants found in wet and dry urban precipitation and in urban waters. Heat releases in urban areas lead to the heat island phenomenon with elevated air temperatures, which then affect local climate and snowmelt.

The soil–water interface is important with respect to soil erosion, leaching of chemicals from soils into water, and on-land disposal of residue from stormwater or wastewater treatment. These interactions may affect both the water and the soil.

Most of the published literature on the effects of urbanization addresses impacts on surface waters, which may include streams, rivers, impoundments, reservoirs, lakes, estuaries and near-shore zones of seas and oceans. The transition from aquatic to terrestrial ecosystems is provided by fresh or saltwater wetlands, which are discussed separately from surface waters.

During the past ten or twenty years, a great deal of attention has been paid to urban groundwater, which is affected by urbanization with respect to both quantity and quality. Depending on local circumstances, urbanization affects aquifers and groundwater tables through either reduced recharge (increased runoff) or increased recharge (leaking water mains, leaking sewers, stormwater infiltration), and pollution by infiltrating effluents or accidental spills.

Finally, changes in the urban environmental compartments greatly affect the urban biota, particularly the fish and wildlife, with respect to health, abundance and biodiversity.

Each of the urban environmental compartments can be subject to various types of impacts which occur in the urban environment in various combinations and magnitude. For convenience, it is useful to discuss these impacts under such headings as physical, chemical and microbiological, but the overall effect is generally caused by the combined impacts of physical, chemical and microbiological nature. Considering the great diversity of topics discussed in this chapter, some guidance to following the

Table 4.1 **Classifications and examples of impacts**

Type of impact	Environmental compartment					
	Atmosphere	Surface waters	Wetlands	Soil	Groundwater	Biota
Physical	Heat island, increased precipitation downwind, dry deposition	Increased surface runoff and flooding, higher water temperature	Changes in water balance	Increased erosion, changes in physical structure	Lower or higher water table	Loss of habitat, benthic organism burial
Chemical	Acid or toxic rain	Pollution of streams and lakes	Pollution	Soil pollution	Dense non-aqueous phase liquid (DNAPL) contamination	Toxic effects, loss of biodiversity
Micro-biological	Small risk of exposure during sludge handling	Faecal pollution of beaches or drinking water sources	Changes in bacterial ecology	Changes in bacterial ecology due to sludge application	Polluted drinking water	Risk of biotic impacts (diseases)
Combined	Smog	Loss of biodiversity, impairment of beneficial uses	Loss of biodiversity, impacts on biota	Landfills	Degraded aquifers	Loss of abundance, loss of biodiversity

material is offered in Table 4.1, which provides classification of urbanization impacts on environmental compartments and illustrates this classification by the examples listed in the table.

4.2 GENERAL CHARACTERIZATION OF URBANIZATION EFFECTS

The process of urbanization changes the landscape as well as material and energy fluxes in the urban areas, thereby affecting the urban environment. Changes in landscape and runoff conveyance are particularly important with respect to surface runoff and its characteristics. Other changes are caused by construction of urban infrastructures, increased water consumption in urban areas, and releases of solids, chemicals, micro-organisms, and waste heat. Water leaves urban areas in the form of urban waste-water effluents (UWWE), which include stormwater, CSOs and municipal wastewaters. Such types of effluents differ in their physical, chemical and microbiological characteristics; consequently, their effects also differ and will be discussed separately within a common framework of impacts. Because of the dynamic nature of UWWE discharges and the associated pollutant levels, loads and effects, the temporal and spatial scales of individual effects are also important (Lijklema et al., 1989). Some effects manifest themselves instantaneously; others may become apparent only after periods of many years. With respect to spatial scales, the magnitude of discharges and the number of outfalls vis-à-vis the type and size of receiving waters are also of great importance. Further discussion of these factors follows.

4.2.1 Increased ground imperviousness

Perhaps the most visible consequence of urbanization is the increase in the extent of the impervious ground cover, which strongly limits the possibility of water infiltration. High imperviousness is particularly noticeable in downtown areas, where its numerical value may exceed 95%. In many countries, the rapid increase in catchment imperviousness is a relatively recent phenomenon; in France, for example, the area of impervious surfaces increased tenfold between 1955 and 1985 (Chocat, 2006).

Increased imperviousness affects runoff in several ways. Firstly, it increases runoff volumes. This effect is often cited when explaining urban floods. However, if the runoff volume increase plays an important role for frequent storm events, or even for the events corresponding to the return periods considered in the design of minor drainage systems (generally about ten years), it is not the most important factor for extreme events.

Indeed, the infiltration capacity of the majority of pervious soils, in the absence of a dense forest cover, or except for sandy grounds, is much lower than the rainfall intensities than can be observed during exceptional rainstorms. Thus, in this type of situation, permeable soils often yield specific runoff volumes (volume of runoff per unit area), which approach those of the impermeable soils. For example, during an extreme flood on the Yzeron River in the Lyon area in April 1989, the runoff coefficient of the rural part of the catchment was estimated at 50% while the corresponding value for the urban part was 60% (Chocat, 2006).

Another significant consequence of increasing ground imperviousness is the lack of recharge of groundwater aquifers (Leopold, 1968). This phenomenon can be accentuated when water is withdrawn from the same aquifer for urban water supply. More importantly, besides the direct effect of depletion of the water resource, the lowering of the water table is likely to cause land subsidence, which in some cases can reach several metres, as reported for example in Mexico City (Figueroa Vega, 1984). Such extreme subsidence then affects the stability of buildings. For example, during the drought spell in France in the early 1990s, payments by the insurance companies for damage to buildings (cracks, fissures, etc.) were ten times higher than the monetary losses in agriculture. However, in certain cases, exfiltration from urban water infrastructures (drinking water mains, sewers and stormwater management measures) can partially compensate for the deficit in rainwater infiltration. For example, in an urban agglomeration of 50 km^2, with imperviousness of 50% and water consumption of 100,000 m^3 per day, water supply distribution network losses (leakage) of 20% are equivalent to a groundwater recharge by infiltration of 300 mm of rainfall per year. Similar values were reported by Lerner (2004).

4.2.2 Changes in runoff conveyance networks

As the urbanizing area develops, there are profound changes in runoff conveyance, as natural channels and streambeds are replaced with artificial channels and sewers.

In general, these changes increase the hydraulic efficiency of runoff conveyance by increasing the speed of runoff. This process starts with overland flow in headwaters of the catchment and progresses to the receiving streams and rivers, which are canalized to increase their hydraulic capacity and protect their beds against erosion. Finally, the general drainage pattern of the catchment is also affected by transportation corridors required in urban areas.

4.2.2.1 Construction of runoff conveyance networks

In urban areas, a natural drainage network, which may be temporary and comprise sinuous waterways partly blocked by vegetation, is likely to be replaced with an artificial conveyance network, which is often oversized in the upstream parts and characterized by a straight layout to limit its length, and laid on significant slopes to decrease drain sizes (and thus reduce costs) and improve its self-cleansing. The same process also takes place in peri-urban areas, with respect to the drainage of soils and the canalization of the brooks, creeks and ditches. This canalization, which is generally presented as an effective means of preventing flooding, often has its origin in the occupation of the flood zone (i.e. a major stream bed) by buildings or roads. However such major stream beds constitute a natural part of the flood plain, and thus play a role in regulating the flows transported to the downstream reaches. The increase in runoff speed and the resulting shortening of the catchment response time contribute to higher runoff peaks through two mechanisms: (a) faster transport processes, and (b) greater intensities of the critical rainfall, which apply to the shortened response times.

4.2.2.2 Canalization of urban streams and rivers

For various reasons, urbanization usually leads to modification of river courses by damming, widening and training. Little brooks are gradually canalized, covered and buried. Most important watercourses are enclosed between high embankments, which completely isolate them from the city. In many cities, after centuries of progressive urbanization, rivers are now regarded only as 'virtual sewers'. The results of this state of affairs are twofold:

- Urban rivers are gradually forgotten by the citizens, who only perceive their harmful effects.
- Urban rivers are enclosed in a too narrow 'corset' and thereby have lost any 'natural' possibility of spilling onto natural flood plains in the case of floods.

The consequences can be catastrophic. The city, which is appropriately protected as long as the water levels remain below the top of the embankments or dams, is suddenly inundated when the flow increases, or these protective structures fail. No longer accustomed to the presence of water, the city then reveals and manifests its increased vulnerability by incurring damage of sensitive equipment located underground (telephone switchboards, electric transformers, pumping stations, etc.), damage to subways and in underground car garages, loss of important supplies of vulnerable goods on ground floors, sweeping away of cars by floodwaters (because of their buoyancy), inexperience of urban dwellers in coping with floods and so on. All the above factors help transform the crisis into a catastrophe.

From an ecological point of view, the anthropogenic changes and river training also have important consequences. A river is indeed a 'living' entity, which must be considered in all its temporal and spatial dimensions. From the spatial point of view the river equilibrium depends on many conditions:

- upstream–downstream continuity (longitudinal dimension)
- habitat diversity (nature of the banks, width of the bed, speed of flow, depth of the river, etc.)
- connections between the major stream bed and hydraulically connected water bodies (lateral dimension)
- flow exchange between rivers and aquifers (vertical dimension).

Construction of embankments and dams, bed dredging, canalization, and construction of new underground structures and foundations, all impoverish the river habitat and decrease its capacity to be regenerated.

The temporal dynamics of rivers must also be considered. The succession of high and low water stages, either episodic or cyclic (temporal dimension), is necessary for river equilibrium. For that reason, the training and regulation of watercourses, building dams to reduce floods and/or to maintain low flows, can be extremely harmful.

4.2.2.3 Interfering transport infrastructures

The third important consequence of urbanization is the construction or expansion of transportation corridors (motorways, railways, etc.). These projects often involve large earthworks; the resulting infrastructures can be either very high compared with the original ground elevations, or very low in the form of deep cuts. As a consequence, these earthworks superimpose a relief on the natural one, which, particularly in flat terrain, can considerably modify the surface runoff and drainage patterns in two ways, both of which contribute to increased flood risk:

- When the linear infrastructure is laid perpendicularly to the slope and the natural direction of flow of water, transportation corridors constitute physical barriers (dams) that force the runoff towards the provided flow openings (culverts), which are generally superimposed on natural and obvious waterways (brooks, main thalwegs of streams, etc.). Earthworks can even, in certain cases, significantly modify the delineation of the catchments themselves.
- When the linear infrastructures are laid in the direction of the slope, they become constructed channels, sometimes with steep slopes, often rectilinear, and always characterized by low flow roughness compared to natural waterways.

4.2.3 Increased water consumption

Growing populations of urban areas and the development of improved sanitation cause an increased consumption of water. This consequence is not always obvious, since urban water demand is usually much lower than agricultural water needs. For economic reasons, the water needed is typically withdrawn from the closest source (a river or aquifer); the pressure on such a water resource can become excessive and result in the lowering of the groundwater table or a severe reduction of low flows. A detailed discussion of this issue was provided in Chapters 2 and 3.

4.2.4 Timescales of urbanization effects

Concerning timescales, two types of UWWE effects are recognized, acute and cumulative (Harremoes, 1988). Acute effects are almost instantaneous and may be caused, for example, by flow (flooding), and discharges of biodegradable matter (impact on dissolved oxygen levels), toxic chemicals (acute toxicity), and faecal bacteria (impacts on recreation). In the case of acute effects, the frequency and duration of occurrence of pollutant levels are important. Transport processes in the receiving waters, including effluent mixing and dispersion, and pollutant decay, are all important and influence the resulting ambient concentrations. The frequency of acute effects is related to the frequency of rain and snowmelt events. However, the duration of such effects exceeds the duration of rainfall and snowmelt events by the duration of the 'wet weather after-effects' (the persistence of the wet-weather disturbance) which varies from a few hours in well-flushed or stable receiving water systems to more than one day in water bodies with limited circulation (Tsanis et al., 1995).

Cumulative effects (sometimes called chronic effects) generally result from a gradual build-up of pollutants or stresses in the receiving waters and manifest themselves only after such accumulations exceed a critical threshold, as may be the case with nutrients and toxicants released from accumulated sediment, or geomorphologic changes in urban streams. For pollutants causing cumulative effects, short timescale dynamics is unimportant and the main interest consists in loads integrated over extended time periods (generally years or decades).

4.2.5 Types of receiving waters and spatial scales

UWWE effects on surface waters also depend on the magnitude of effluent discharges and the type and physical characteristics of the receiving waters. All receiving water bodies can cope with some input loads without a serious impairment of their integrity, but problems arise when this capacity is exceeded. With reference to UWWE discharges, the receiving surface waters range from creeks and streams, rivers, lakes or reservoirs of various sizes, to estuaries and oceans.

The UWWE effects are most serious in small urban creeks with wet-weather inflows vastly exceeding the creek flow and minimal dilution of effluent discharges. Small streams can be severely impacted by cumulative effects of elevated discharges, and incoming discharges of chemicals, pathogens and heat. The morphology of such streams may change dramatically, and these changes contribute to physical habitat destruction. In rivers, the mixing and dispersion of UWWE pollutants are important processes that reduce pollutant concentrations outside the mixing zones. Acute effects are generally less common than in streams/creeks because of large input dilution and the riverine self-purification capacity. Typical urban runoff flows may be too small to affect the stream geomorphology, which is governed by the streamflow regime.

UWWE effects on lakes and reservoirs depend on the size of the water bodies. Small impoundments in urban areas (e.g. stormwater ponds) are most heavily affected, particularly by faecal bacteria, nutrients and contaminated sediment. The large influx of sediments also destroys the physical habitat. In the case of large lakes (e.g. the Great

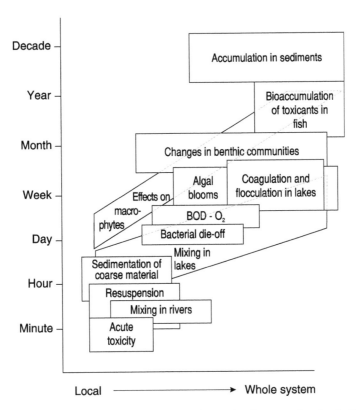

Figure 4.1 Relation between the rates of processes and spatial and temporal scale of effects

Source: after Lijklema et al., 1989.

Lakes in Canada and the United States), UWWE discharges typically affect only the near-shore waters in the vicinity of urban areas and their sewer outfalls. Finally, potential effects of UWWE on ocean waters are minimal; however, municipal effluents, in general, do affect harbours, estuaries and coastal waters.

The characteristics of the urban area and the receiving waters determine the spatial scales of UWWE effects. Stormwater outfalls are dispersed in large numbers throughout urban areas, with hundreds or thousands of outfalls. CSO outfall points are consolidated to fewer locations. Finally, sewage effluents, which are also called point sources of pollution, are discharged typically at one or a few locations along the river, usually downstream of the urban area, or via outfalls which may extend deep into a lake or ocean. Such arrangements reduce sewage effluent impacts on urban waters. On the other hand, pollutant transport in the receiving waters further increases the spatial extent of UWWE effects.

Spatial and temporal scales of various processes in receiving waters are shown in Figure 4.1 (Lijklema et al., 1989).

4.3 URBANIZATION EFFECTS ON THE ATMOSPHERE

Urban areas and land use activities have many types of effects on the atmosphere. A full discussion of such effects is outside the scope of this report, which focuses on water issues, and consequently only the issues relevant to urban water are briefly introduced. The most obvious connections between the urban atmosphere and water include elevated temperatures that affect urban precipitation and snowmelt, and air pollution contributing to wet and dry deposition in urban areas, pollution of precipitation, and increased precipitation due to the presence of condensation nuclei.

4.3.1 Thermal effects (urban heat island phenomenon)

Urban areas are known to attain higher air temperatures than the surrounding rural areas, because of differences in thermal balances. Higher temperatures in cities (by 4–6°C, Geiger et al., 1987) result from several factors, including: (a) net solar energy gains, which are not moderated by evaporative cooling (due to the relative lack of water surfaces or vegetation subject to evapotranspiration), (b) waste heat from buildings and means of transportation (this may represent up to one-third of solar energy input), and (c) the 'canyon structure' of tall buildings that traps solar energy and reduces infrared heat losses. The latter phenomenon may be somewhat mitigated by strong winds, and may increase cloudiness and precipitation in the city, as a thermal circulation sets up between the city and surrounding region (Myer, 1991).

4.3.2 Urban air pollution

Air transport of pollutants through the atmosphere occurs as either short-range or long-range transport, with the former being common and intense in urban areas. Air pollution originates in urban and industrial areas from gaseous emissions, volatilization of toxic compounds from water and soil, and particles from land use activities and wind erosion. The transport of pollutants in the atmosphere, with respect to the direction and distance travelled, and concentration levels can be described as a pollution plume, with the highest concentrations near the emission source. The dispersion of pollutants in the atmosphere depends to a large extent on the height of the mixing layer, wind velocity and the atmospheric temperature. Only certain pollutants are routinely measured in the atmosphere, including sulphur dioxide, nitrogen oxides, suspended particles, carbon monoxide and ozone. The acid rain precursors – oxides of nitrogen and sulphur – and particles are relatively easily transferred to water and soils; other constituents, because of their solubility and instability, may not be readily transferable to other environmental media. Typical sources of air pollution include traffic, heating, industrial operations, paints, gasoline evaporation and gas leaks. A detailed assessment of air pollution in twenty megacities of the world was provided by UNEP and WHO (1992).

Air contaminants contribute to pollution of urban precipitation in the form of acid or polluted rain. SO_x and NO_x combine with atmospheric water to produce H_2SO_4 and HNO_3 acids. Various forms of precipitation, such as rain, snow, dew, fog, sleet or hail, transport these acids into soils and water bodies. Acid rain can be generated and deposited locally, but it can also be transported hundreds or thousands of kilometres by wind, and affect large areas. Acid rain in urban areas corrodes metallic surfaces and damages urban infrastructure, buildings, historical monuments, and plants, and also affects water bodies and soils with low alkalinity or capacity to control pH changes (Henry and Heinke, 1989). In vegetated areas it weakens trees and other plants, and in soils it limits the nutrient availability to plants and extracts aluminium from soils, which impairs soil structure and permeability. Acid rain may also leach metals from soils, and such dissolved metals may be toxic to aquatic biota. Acidic deposition accumulates in snowpacks and is suddenly released during the final snowmelt, causing an acidic shock in receiving streams that results in fish kills.

Besides inorganic acids, organic acids, which are toxic to biota at certain levels, are also commonly found in the air and marginally contribute to the acid rain phenomenon. Of these, formic, acetic and propionic acids are more common in water, as they are highly soluble. Hydrocarbons are also transferred from the air to water and soils by atmospheric deposition. They originate from combustion or volatilization of fuels, and the use of paints and solvents. Observations in Mexico City indicate that annual air emissions include 20,000 tonnes of particles smaller than 10 microns, 22,500 tonnes of SO_2, 1,800,000 tonnes of CO_2, 206,000 tonnes of NO_x and 465,000 tonnes of hydrocarbons (Molina and Molina, 2002). Nitrogen compounds in the atmosphere contribute to the eutrophication of lakes as well as to increasing nitrate concentrations in groundwater.

Air in cities contains solid particles of diverse sizes and origins. Such particles are scavenged by rain and transported into soils and surface waters. The composition of particles is of great concern, because they frequently absorb metals, toxic organic compounds and gases. For example, fine particles produced by diesel vehicles have been shown to contain 1,3-benzene and butadiene, which are both well-known carcinogenic compounds (Fernandez-Breaumontz, 1988). Besides water quality effects (and human health effects), the presence of particulates in the atmosphere contributes to increased precipitation, because such particles serve as condensation nuclei in formation of rainfall droplet. Increased precipitation (by 5–30%) downwind from urban areas was reported by Geiger et al. (1987).

4.3.3 Combined impacts

From the human health point of view, a particularly dangerous form of air pollution is smog, which is a noxious mixture of air pollutants (fine particulate and gases) held in stagnant masses of air over urban areas, particularly during thermal inversion. This phenomenon was first observed – and the term smog coined – in London, UK, where a combination of fog and smoke created serious air pollution problems that resulted in thousands of additional deaths during smog spells. The two key components of smog are airborne particles and ground-level ozone. Airborne particles are solids or minute droplets of liquid that are suspended in the air for days or even weeks. Ground-level ozone is produced by photochemical reaction between nitrogen oxides (NO_x) and volatile organic compounds (VOC) (UNEP and WHO, 1992).

4.4 URBANIZATION EFFECTS ON SURFACE WATERS

Urbanization strongly impacts on surface waters, particularly in the case of streams, small rivers, impoundments, lakes, estuaries and coastal waters. The physical setting and explanations of such effects were introduced in Section 4.2; a more detailed discussion of various types of effects follows.

4.4.1 Physical effects

4.4.1.1 Urbanization effects on flows

The process of urbanization results in high catchment imperviousness (increased volume of runoff), fast runoff (increased speed of runoff) and quick catchment response to critical rainfall of reduced duration, all of which contribute to increased runoff flows, as demonstrated in Figures 2.2 and 2.3 in Chapter 2, which display attributes of pre-development and post-development runoff hydrographs.

While the first two reasons for runoff peak increase after urbanization (i.e. increased volume and speed of runoff) are well understood, the third reason, the reduced catchment response time, making the catchment more sensitive (responsive) to rain events of shorter duration and higher intensity, and therefore producing higher specific runoff rates, is less obvious and deserves further discussion. Theoretically, for a given homogeneous catchment, the critical rainfall producing the greatest runoff is that whose duration, t_d, is equal to the time of concentration of the catchment, t_c. Indeed, if the duration is shorter ($t_d < t_c$), the whole catchment area does not contribute fully at the same time to the flow at the catchment outlet. On the other hand, for a given return period, the shorter the rain duration, the higher its average intensity. For that reason, shorter response times result in shorter times of concentration, and greater intensities of the critical rainfall. This phenomenon undoubtedly plays the most important role among the factors discussed. Numerical simulations show that a reduction in the catchment response time by 20–50% leads to an increase in corresponding peak flows by a factor ranging from five to fifty-fold.

Water balances of impoundments and lakes receiving urban runoff are also affected, with generally higher inflows during the wet seasons and lower inflows in summer.

4.4.1.2 Urbanization effects on sediment regime: erosion and siltation

Among the effects of increased runoff flows and their durations are secondary effects on sediment erosion, transport, increased concentrations of suspended solids and sediment deposition (siltation) in slowly moving stream reaches. Soil erosion is intensified in urbanizing areas as a result of two factors: the stripping of natural protective vegetative cover from the soil surface during construction, and increased runoff flows, which cause sheet erosion, scouring in unlined channels and transport of eroded material to the downstream areas (Horner et al., 1994). Wolman and Schick (1962) reported that sediment yields from natural catchments were as low as $100\,t/km^2/year$, but increased more than 100 times during urbanization. After completion of the urban development and the establishment and consolidation of surface covers, the sediment yields drop to the predevelopment, or even lower, values (Marsalek, 1992). Thus, excessive soil erosion in urbanizing areas is a transient process, which should be mitigated by implementation of erosion and sediment control programmes.

Excessive erosion causes ecological damage by sweeping away habitats (Borchardt and Statzner, 1990) and expanding stream channels (in both width and depth) either gradually, or as a result of a single severe storm that results in rapid downcutting or channel incision (Booth, 1990; Urbonas and Benik, 1995). Eroded soils contribute to increased concentrations of suspended solids, which cause a number of direct and indirect environmental effects. These effects include those associated with reduced sunlight penetration (interference with photosynthesis); blanketing of gravel substrates where fish spawn and rear their young, and where algal and invertebrate food sources live; filling up pools where fish feed, take refuge from predators and rest; damaging fish gills and other sensitive tissues; reducing visibility for catching food and avoiding predators; transporting various pollutants; and, contributing to loss of riparian vegetation (with the concomitant loss of shade and refuge) and large woody debris forming a part of aquatic habitat (Horner et al., 1994). Another cause of stream siltation is the large load of suspended solids carried by treated and untreated sewage effluents and discharged to the receiving waters.

Sediment conveyed by urban runoff is deposited in receiving impoundments and lakes, where it causes similar effects as in streams, particularly with respect to siltation and increased concentrations of suspended solids in the water column.

Erosion and siltation impacts, including adverse morphological changes in urban streams (Schueler, 1987), can manifest themselves on various timescales; a single large rainfall/runoff event can cause significant impacts, but generally long-term impacts are more important. Ecological impacts include those related to critical species and dispersal and migration, and practically all beneficial water uses are affected (water supply, bathing, recreation, fishing, industrial water supply and irrigation) (Lijklema et al., 1993).

4.4.1.3 Modification of the thermal regime of receiving waters

The urban environment contains numerous sources of heat, which increase the temperature of surface runoff. This warming of runoff is particularly strong during the summer months, when rainwater comes into contact with hot impervious surfaces (pavements, roofs) or is exposed to solar radiation in storage facilities (Van Buren et al., 2000), and stream discharges are low. The resulting stormwater temperatures may exceed those in the receiving waters by up to 10°C (Schueler, 1987) and contribute to long-term changes in the receiving water temperature as the development of the basin progresses. Additional waste heat is conveyed by WWTP effluents and cooling waters (e.g. from chemical industries and power plants), which are generally discharged at a single point.

Aquatic organisms have characteristic temperature preferences and tolerance limits. As catchment development progresses and thermal enhancement takes place, the original cold-water fishery may be succeeded by a warm-water fishery, cold-water algal species (mainly diatoms) may be replaced by warm-water filamentous green and blue-green species, and cold-water invertebrates (where they exist) are also adversely affected (Galli, 1991). Ecological impacts of thermal enhancement include those related to energy dynamics, food web, genetic diversity, and dispersal and migration. The most heavily affected beneficial water use is fishing (Lijklema et al., 1993).

4.4.1.4 Density stratification of receiving water bodies

Density stratification of urban receiving waters may be caused either by dissolved solids or by temperature. Large quantities of total dissolved solids may be found in urban storm-water in cold climate regions, where large quantities of chloride are conveyed by runoff and snowmelt as a result of road salting (Marsalek, 2003b). Such loads then lead to densimetric stratification of urban lakes and impoundments (Judd, 1970; Marsalek, P.M. 1997), with a concomitant impediment of vertical mixing and oxygenation of the bottom layers of such water bodies. Stratification can be also of thermal origin, as observed in natural water bodies and further described in Section 4.4.5.2.

Furthermore, high concentrations of chloride contribute to increased mobility of heavy metals (Novotny et al., 1998) and occurrence of toxic effects (Rokosh et al., 1997; Marsalek et al., 1999a), with the resulting loss of biodiversity (Crowther and Hynes, 1977). The ecological impacts include those on the food web, genetic diversity and ecosystem development. The affected beneficial water uses include water supply, fishing and irrigation (Lijklema et al., 1993).

4.4.1.5 Combined physical effects

Throughout the world, very few river basins remain unaffected by anthropogenic influences. The general deforestation of land, changes in terrain slopes to allow agricultural cultivation, and damming of creeks and rivers for energy production, irrigation or navigation, have strongly modified the hydrological behaviour of the affected streams and rivers. Thus, the popular references to achieving a 'natural' state (or renaturalization) of catchments, in a vast majority of cases, are somewhat Utopian. In practice, the actual reference to the 'natural' state is of a historical character and often corresponds to an equilibrium state, actual or hypothetical, which was reached in the catchment in a not-too-distant past, so that the collective memory preserves its trace. All water bodies, and particularly those in urban areas, are always in a state of permanent evolution. Thus, the concept of equilibrium indicated here does not correspond to a stable and invariable state, but to a certain stage of slow and partly controlled evolution, striving to preserve the physical (water resource, floods, solid transport), biological (nature of the ecosystems) and sociological (beneficial uses of the water body) balances. Yet urbanization still affects the behaviour of catchments, and it is necessary to describe the different kinds of effects that can be observed in order to be able to mitigate the most disastrous ones. Further complexity and uncertainty may be introduced into this analysis by climate change, which may affect all aspects of the hydrological cycle.

The various types of impacts of urbanization on the hydrological cycle are not independent; in fact, they act synergistically, as shown in Figure 4.2. The ultimate result is

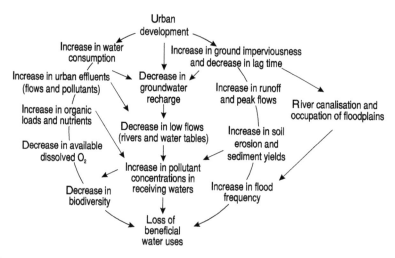

Figure 4.2 Impacts of urbanization on the aquatic environment

Source: after Chocat, 1997.

a somewhat paradoxical situation. In urban areas, where the requirements for water resources, with respect to water quantity, quality and security, are the strongest, the risks of flooding are the highest and the aquatic environment and its ecosystems are the most degraded.

4.4.2 Chemical effects

In this section, environmental and ecological problems caused by the pollution of urban wastewater effluents are examined. The individual types of effluents – stormwater, CSOs and municipal wastewater – were characterized in Chapter 3 (Sections 3.3.2.1, 3.3.3.1 and 3.4.5, respectively). The discussion presented in this chapter focuses on their chemical effects in receiving waters.

4.4.2.1 Dissolved oxygen reduction

Reductions in dissolved oxygen (DO) and the concomitant biomass accumulation are typically caused by discharges of oxygen-demanding substances, characterized by bio-chemical oxygen demand (BOD), chemical oxygen demand (COD) and ammonia. Oxygen-demanding substances are conveyed in relatively high concentrations by WWTP effluents (Chambers et al., 1997) and CSOs (Harremoes, 1988); stormwater sources are much less important. However, DO depletion can occur in summer months in shallow stormwater ponds, or in ice-covered stormwater ponds during the winter months (Marsalek, P.M. 1997). Environmental effects occur on two timescales: short-term effects are caused by dissolved BOD/COD and ammonia, and intermediate-term effects may be caused by sediment oxygen demand (Hvitved-Jacobsen, 1982). These impacts have effects on both the ecology and water uses; ecological effects include those on biodiversity and critical species; the affected water uses include water supply, bathing, fishing and industrial water supply (Lijklema et al., 1993).

4.4.2.2 Nutrient enrichment and eutrophication

Nutrient enrichment and eutrophication of receiving waters is typically caused by nitro-gen and phosphorus loads discharged to receiving waters with stormwater, CSOs and WWTP effluents. Among these sources, WWTP effluents, particularly if inadequately treated, are the most important. Excess nutrients can stimulate the growth of primary producers (algae and rooted aquatic plants) to levels that impair the ecosystem by changes in energy dynamics and food web structure, changes in the composition of algal community from one-celled diatoms to filamentous green forms, followed by blue-green forms, changes in habitat and loss of species (Chambers et al., 1997). Eutrophication degrades lake ecosystems in a number of ways, including reduced food supplies to herbi-vores, reduced water clarity, and at the end of the bloom, algal decomposition which causes high oxygen demands that lead to oxygen deficiency, particularly in the bottom layers. These effects typically manifest themselves over longer time scales (seasonal or even longer) (Harremoes, 1988). The affected beneficial water uses include water supply, bathing, recreation, fishing, industrial water supply and irrigation (Lijklema et al., 1993). The prevention of urban lake or reservoir eutrophication usually requires control of nutrient sources, including WWTP effluents, stormwater and CSOs (Schueler, 1987), and upstream agricultural sources (where they exist). The vertical variation of water quality that may occur in lakes illustrates the importance of having a water intake tower with ports at various depths so that the water supply can be drawn from the most advantageous level in the water quality profile (Hammer and Hammer, 2003).

4.4.2.3 Toxicity

All three types of UWWEs may cause toxic effects in receiving waters by causing elevated levels of ammonia, total residual chlorine, cyanide, sulphides, phenols, surfactants, chlorides, metals, and trace organic contaminants (e.g. organochlorine pesticides). Ambient conditions in the receiving waters, such as temperature, pH, hardness, alkalinity and DO may modify the toxicity of chemicals, and interactions of chemicals in receiving waters may increase the risk of toxicity. With reference to WWTP effluents, the extensive literature on this topic indicates that toxicity is usually attributed to ammonia, total residual chlorine (where used in disinfection), cyanide, sulphides, phenols, surfactants and some heavy metals (Cu, Zn, Cr and Ni). Ambient factors and pollutant interactions modify toxicity in receiving waters and impede generalizations (Chambers et al., 1997).

The understanding of toxicity of CSOs and stormwater is incomplete, but generally their toxicity is attributed to ammonia, toxic metals, hydrocarbons (particularly polycyclic aromatic hydrocarbons, PAHs), and pesticides (Hall and Anderson, 1988; Dutka et al., 1994a, 1994b). The toxicity of urban runoff is measured by bioassays, but conjunctive determination of causes (i.e. pollutants and their forms) is lagging behind (Marsalek et al., 1999a, 1999b). Results from selected studies are summarized below.

Hall and Anderson (1988) used *Daphnia pulex* to investigate stormwater toxicity caused by trace metals in runoff from various urban land uses. Toxicity varied greatly and was attributed to several metals (copper, lead, iron and zinc). Dutka et al. (1994a, 1994b) reported toxicity assessment of water and sediment from four stormwater ponds in the Toronto area. A large number of samples of pore water and solvent extracts tested positive for the presence of promutagens; a limited search for sources of toxicity pointed towards ammonia and pesticides. Marsalek et al. (1999a, 1999b) studied urban runoff toxicity at fourteen urban sites, including two sites receiving runoff from major multi-lane divided highways (>100,000 vehicles/day). About two-fifths of all data did not show any toxic responses, one-fifth indicated severe toxicity, one-fifth confirmed toxicity, and one-fifth potential toxicity. The highest occurrence of severe toxicity was found at sites receiving highway runoff.

In comparisons of CSOs and stormwater toxicity (Marsalek and Rochfort, 1999), CSOs displayed lower acute toxicity (only 7% of samples were moderately toxic), and none of the samples was severely toxic. The frequency of genotoxicity detection in CSOs was higher than acute toxicity detection (15% of samples were at least moderately genotoxic) and up to two-thirds of all CSO samples showed chronic toxic effects, depending on the toxicity test applied.

In receiving waters, the observed effects of stormwater and CSO toxicity may be offset by mixing and dilution with less polluted ambient water. Toxicity measurements were found effective in screening and comparing sources of toxicants, but their effectiveness in prediction of control performance and the assessment of chronic toxicity has not yet been demonstrated (Marsalek et al., 1999b).

Ecological impacts of ammonia and trace organic contaminants (toxicants) include those on the food web, biodiversity and critical species; in the case of metals, such a list could be further expanded for ecosystem development. In the short term, the only beneficial water use significantly affected is fishing (Lijklema et al., 1993); in the long term, the receiving water ecosystem is downgraded.

4.4.3 Microbiological effects

Microbiological contamination of urban waters is one of the most severe impacts of urbanization, because it adversely affects the health of urban dwellers. Such a contamination is directly related to the level of solid waste management and sanitation in urban areas. Furthermore, in densely populated areas of the developing world it is difficult to find alternative solid waste disposal sites. Existing waste disposal sites which are no longer in use may end up as informal settlements. Leachates from landfill sites represent another source of contamination.

In developed countries, this level of waste management is fairly advanced and the associated effects on human health are limited. However, the situation is very different in developing countries, where large populations lack access to improved sanitation. The main sources of microbial contamination of urban waters are human wastes. The microbial agents causing waterborne illness were described, for example, by Leclerc et al. (2002).

4.4.3.1 Waterborne pathogens

There are four groups of organisms that affect human health and are commonly found in water: viruses, bacteria, protozoa and helminths. The first three are microscopic, while the latter measure from millimetres to metres, but in water travel as microscopic eggs. Table 4.2 contains a list of pathogenic bacteria which are likely to be found in sewage. Their presence and concentrations depend basically on local conditions.

Viruses exist in different forms and sizes (from 0.01 to 0.3 μm) and reproduce only within infected cells. There are more than 140 enteric types of viruses that cause infections or diseases. Unlike bacteria, pathogenic viruses are not found in the faeces of healthy humans, but only in those of infected individuals. The enteroviruses of greatest concern are polio, echo, coxsackie, Norwalk, rotavirus, reovirus, calicivirus, adenovirus and hepatitis A. Viruses are more resistant to disinfection and environmental conditions than most bacteria and are very difficult to detect using conventional laboratory techniques. Rotavirus is the most important cause of infantile gastroenteritis, causing 500–1,000 million annual episodes of diarrhoea in children under five years in Africa, Asia and Latin America, and up to 3.5 million deaths (Jawetz et al., 1996).

Bacteria are unicellular micro-organisms measuring between 0.1 and 7 μm and occurring in diverse forms – spherical, oval, helicoid, filamentous or rods. They are ubiquitous in the environment and occur in many varieties, some of which are innocuous to man. Some of them colonize the human intestines, like faecal coliforms, and are evacuated in human faeces in great quantities (more than 10^{12} per gram). The bacteria that pose the greatest health risk are the enteric bacteria that live in or can inhabit human intestines. Such bacteria are adapted to living in an environment rich in organic matter at 37°C, and consequently have difficulty surviving in other environmental conditions. The main pathogenic bacteria transmitted through water are listed in Table 4.2.

Protozoa are parasites often associated with diarrhoea. They are unicellular, measure between 2 to 60 μm and are found in two forms: trofozoite and cyst. Infection is

Table 4.2 Main pathogenic bacteria transmitted through water

Bacterium	Characteristics	Illness
Escherichia coli	Some breeds are pathogens.	Intestinal infections.
Campylobacter jejuni	The main cause of gastroenteritis in Europe (rather than Salmonella); originates from non-chlorinated water sources.	Diarrhoea in humans and animals (more frequent); usually affects children and young people; incubation period 2–5 days.
Salmonella	Widespread; one of the most important pathogens affecting both humans and animals due to amount of serotypes that exist; very common in water and food in developing countries.	Salmonellosis or typhoid fever causes acute gastroenteritis with diarrhoea, abdominal cramps, fever, nausea, vomiting, headaches and, in severe cases, death; frequency varies annually and differs from one country to another; the infective dose varies from 10^5–10^8 although it can be produced in low immunity individuals at 10^2 for Salmonella typhi.
Shigella	Does not survive well in the environment; a large amount of serotypes exist (more than 40), but S. sonnei and S. flexeneri account for 90% of the isolates from raw wastewater.	Bacillary Shigellosis or dysentery causes fever, nausea, vomiting, abdominal pain, migraine and faeces emission with blood and mucous; highly infectious; more virulent in old people and chil dren; the infective dose is 10^3 microorganisms.
Mycobacterium tuberculosis	Causes diseases in people who swim in contaminated water.	Gastrointestinal alterations.
Vibrio cholerae	Normally present in the aquatic environment; its presence depends on water temperature and salinity. Vibrio cholerae is not very common in developed countries but frequent in developing countries; humans are the only well-known hosts and the most frequent way of transmission is the ingestion of polluted water or of produce irrigated by polluted water.	Gastroenteritis; usually affects children; causes very abundant liquid diarrhoea, with important hydroelectrolytic losses and severe dehydration, associated with vomiting.
Helicobacter pylori	The means of transmission are not well known; unsanitary conditions and consumption of polluted vegetables are possible means; more sensitive to chlorine than faecal coliforms.	Causes gastritis, duodenal ulcer (peptic), gastric ulcers and carcinoma.

Source: after Craun, 1984; Sansonetti, 1991; Thomas et al., 1992; Hopkins et al., 1993; Lima and Lima, 1993; Nachamkin, 1993; Jawetz et al., 1996; Johnson et al., 1997.

acquired by ingestion of the mature cyst, which is resistant to gastric juice. In the small intestine the cyst transforms itself into a trofozoite and settles down. Trofozoites can once again form cysts and be eliminated in human faeces whether or not the person displays the symptoms. Like viruses, protozoa do not reproduce in the environment; nevertheless, they are able to survive in it for weeks, months or years, depending on the environmental conditions (Bausum et al., 1983). Protozoa are transmitted via polluted food or water. Table 4.3 lists the main characteristics of diverse protozoa.

Helminths are pluricellular organisms. Generally, free-life larvae are not pathogenic, but those in wastewater are pathogenic and are associated with suspended solids. Helminths contribute to poor nutrition, anaemia and delay growth. There are different species of helminth ova, whose relative frequency depends on regional conditions, with *Ascaris* commonly found in wastewaters. For example, in Mexico City's wastewater, 90% of helminth ova were attributed to *Ascaris* (Jiménez and Chávez, 1998).

The ova measure between 20 and 100 μm and are resistant to diverse environmental conditions as well as conventional disinfectants (chlorine, ozone and UV light). They can be removed from wastewater by sedimentation, coagulation-flocculation

Table 4.3 Main protozoa transmitted through water

Protozoan	Characteristics	Illness
Entamoeba histolytica	10% of the world population suffers from amoebas (500 million infected people); 40 to 50 million cases and up to 100,000 deaths occur per year (the second cause of mortality by parasites after malaria); 96% of all cases occur in poor countries, particularly on the Indian subcontinent, in Western Africa, the Far East and Central America.	Invade the large intestine, occasionally penetrate the intestinal mucous and lodge in other organs; responsible for amoebic and hepatic dysentry. The prevalence depends on cultural habits, age, sanitary conditions and socio-economic conditions. In developed countries, this disease occurs mainly among immigrants.
Cryptosporidium	Widely distributed in the environment; infects farm animals and pets. It was recently (1976) discovered to be a human pathogen. Infected people carry it for life and can be reinfected.	Cryptosporidiosis causes stomach cramps, nausea, dehydration and headaches. The infective dose is 1–10 cysts. Different population segments and different cultures react in different ways.
Giardia	Giardiasis is endemic, with 10% prevalence in developed countries and 20% in developing ones; the total number of cases is 1,100 million per year, 87% of which are in developing countries. This incidence has been increasing in recent years (WHO, 1997); water may not be the main transmission mechanism.	Very liquid, odorous and explosive diarrhoea, stomach and intestine gases, nausea and loss of appetite. The incubation period is 1–4 weeks. It particularly affects undernourished children under five.

Source: after Salas et al., 1990; Gray, 1994; Goldstein et al., 1996; Tellez et al., 1997; WHO, 1997; Cifuentes et al., 2002.

(Jiménez et al., 2001), stabilization lagoons (WHO, 1989), wetlands and filtration (Jiménez et al., 2001). They are also inactivated by high temperatures, and dry and acid conditions (Barrios-Pérez, 2003).

Helminthiasis affects 25% to 33% of the population in developing countries (Wani and Chrungoo, 1992; Bratton and Nesse, 1993), whereas in developed countries less than 1.5% are affected (WHO, 1997). Ascariasis is endemic in Africa, Central and South America, Asia and the Far East, especially in regions where poverty, congested living conditions and bad sanitary conditions predominate. In such situations, incidence of helminths can be found in 90% of the population (Schulman, 1987; Bratton and Nesse, 1993). In the United States, ascariasis is common, but of the 4 million infected people, most are immigrants (Bratton and Nesse, 1993). In this sector, the prevalence rate is 20–60% (Salas et al., 1990). Basic statistics concerning helminthiasis are summarized in Table 4.4.

In developed countries, microbiological pollution effects on human health and biomass are primarily associated with CSOs, and to a lesser degree with stormwater. The effects on public health are related to swimming beaches; the effects on biomass include contamination of shellfish and closure of harvesting areas.

Stormwater, CSOs and non-disinfected (or poorly-disinfected) WWTP effluents convey high loads of faecal bacteria, which are typically described by concentrations and fluxes of indicator bacteria, such as *Escherichia coli* (EC). Various public health authorities have established recreational water quality guidelines of two types: (a) regulatory schemes based on compliance with indicator organism limits (IOLs), and (b) a risk-based approach (WHO, 2003).

In guidelines based on compliance with IOLs, public health authorities establish the tolerable levels of health risk for swimming, and when the associated IOLs are exceeded, beaches are either closed to public swimming, or posted as 'unsafe'. Where such limitations on use of recreational waters are deemed excessive, remedial programmes designed to improve water quality are developed and implemented. For example, the current Province of Ontario (Canada) limit, calculated as a geometric mean of no less than five samples, is 100 EC/100 mL. While the determination of microbial pollution in the receiving waters is a routine task, the potential public health risks are not well understood (lack of epidemiological data). Furthermore, these effects manifest themselves instantaneously, though their measurement (involving laboratory incubation) introduces time delays into the process of detection, and the assessment of compliance with the existing water quality guidelines is defined as a geometric average of a number of individual measurements collected over extended time periods, which contradicts the instantaneous nature of impacts of this parameter.

Table 4.4 Infected people and annual cases and deaths caused by helminthiasis

Parasitosis	Number infected (million)	Annual cases (million)	Annual deaths (thousand)
Amibiasis	500	40–50	40–100
Giardiasis	200	0.5	–
Ascariasis	800–1,000	1	20
Uncinariasis	700–900	1.5	50–60
Tricocefalosis	500	0.1	–

Source: after Salas et al., 1990; WHO, 1997.

The risk-based approach forms a foundation for the WHO guidelines for safe recreational environments (WHO, 2003) and comprises two steps: (a) risk assessment for swimming in recreational waters, and (b) risk management by establishing objectives for water quality corresponding to a tolerable risk of illness and associated management measures. Furthermore, the guidelines accept the fact that in some areas microbial water quality may be strongly influenced by short periods of high faecal pollution caused by rainfall/runoff events (Ashbolt and Bruno, 2003; David and Matos, 2005; Haramoto et al., 2006).

Both approaches require the assessment of faecal contamination of receiving waters by sanitary inspection and indicator organism counts. The sanitary inspection should identify all sources of faecal pollution, including human faecal pollution conveyed by urban effluents, urban wildlife (particularly birds), domestic pet populations (particularly dogs), beach sand, lack of sanitation, poor solid waste management, land wash, and growth of bacteria in nutrient-rich stagnant waters (WHO, 2003). Land-based sources contribute via discharges of urban stormwater, CSOs and WWTP effluents; birds and pets may also contribute directly to the waters and beach sand. Some land-based sources are activated during wet weather; sand and sediment sources in receiving waters (i.e. beach sand) are activated by waves and currents (LeFevre and Lewis, 2003; Kinzelman et al., 2004). Indicator organism counts are established by field monitoring (WHO, 2003), which can be aided by microbial source tracking (MST) to identify the origin of indicator bacteria with respect to such sources as humans, wildlife (particularly birds), domestic animals and pets (dogs and cats). Even though the underpinning science is still in the developmental stage, MST is applied in studies of recreational waters and provides an additional line of evidence in identification of bacterial sources and the selection of appropriate management (risk control) measures (Edge and Schaefer, 2006).

As typical concentrations of *E. coli* in CSOs may reach up to 10^6/100 mL, and up to 10^5/100 mL in stormwater, these sources often cause bacterial contamination that exceeds the recreational water quality guidelines in receiving waters. Such exceedances occur during wet weather and usually persist for a significant time period afterwards (depending on bacteria die-off and their transport in the receiving waters), often lasting 24–48 hours after the end of the storm (Tsanis et al., 1995). Thus, the beaches affected by urban stormwater and CSOs may have to be closed for periods of several days – comprising the duration of the storm and the persistence of after-effects.

Many beaches in urban areas are frequently closed during and immediately after rainfall events, because of faecal bacteria contamination caused by stormwater and CSO discharges (Dutka and Marsalek, 1993; Marsalek et al., 1994; Tsanis et al., 1995). The reported sources of faecal bacteria in CSOs and stormwater include sanitary sewage, pet populations, urban wildlife (particularly birds), cross-connections between storm and sanitary sewers, lack of sanitation, deficient solid waste collection and disposal, accumulation of sediments in sewers and receiving waters, rodent habitation in sewers, land wash and the growth of bacteria in nutrient-rich standing waters (Olivieri et al., 1989).

4.4.3.2 Indicators of microbiological pollution

Because of inherent difficulties with measuring pathogens, microbiological indicators are generally used instead. An ideal microbiological indicator should: (a) be present

exclusively when there is faecal contamination, (b) have equal or greater capacity to survive than the pathogens, (c) not easily reproduce in the environment, and (d) be measurable in environmental samples. Even though no indicator fulfils all of these conditions
perfectly, faecal coliforms or *E. coli* are currently used as indicators in freshwater, and
enterococci in marine waters. However, these micro-organisms are not adequate indicators of the presence of viruses, protozoan and helminth ova, but only of bacteria. In countries with widespread waterborne diseases, selection of an indicator is a complicated
process, and generally it is recommended that a combination of organisms should be used.

There are a great many different viruses and thus it is impossible to measure them
all. A virus indicator would be helpful. The use of bacteriophage viruses as indicators
was proposed, because they can be easily detected. These, however, are viruses that
infect bacteria but have not been related to human viruses. Attempts to use the col-
iphage group have been made, because of their relatively high concentrations in wastewater and simple and fast (24 h) detection technique, but they do not mimic
enterovirus behaviour and further research is needed.

Protozoon cysts and helminth ova are very resistant to disinfection and diverse environmental conditions. Currently, any protozoon is considered a good indicator, but the
applicable analytical procedures are very complex. For helminths, *Ascaris* eggs are generally measured; however, in this case, given the type of technique (visual counting), it does
not make much sense to speak of an indicator but rather to measure all the species of concern in a certain region, since it entails virtually the same effort, time, cost and training.

Ecological impacts of microbiological pollution include those on energy dynamics,
the food web and ecosystem development. The affected beneficial water uses include
water supply, bathing and fishing (Lijklema et al., 1993).

4.4.4 Combined effects on surface waters

Urban wastewater effluents, including WWTP effluents, stormwater and CSO discharges,
often in combination with other stressors, cause numerous biological effects through a
combination of five factors: (a) flow regime changes, (b) impairment of habitat structure, (c) biotic interactions, (d) changes in energy (food) sources, and (e) chemical variables (water pollution). The effects of these factors are measured by the biological
community performance, as shown schematically in Figure 4.3.

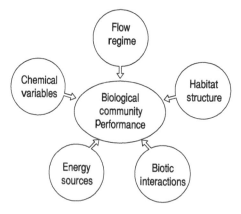

Figure 4.3 **Combined effects on biological community performance**

4.4.5 Examples of urbanization effects on specific types of receiving waters

The effects of urbanization on streams, rivers, lakes and reservoirs are presented in this section.

4.4.5.1 Rivers

Rivers have highly diverse characteristics, depending on the catchment characteristics, climate, pollution sources, flow velocities and other factors. Generally, river flows are highly variable in time (annual and cyclic variation) and space, in terms of topography, climatic and basin conditions. In general, strong vertical mixing occurs in rivers due to currents and turbulence, but full lateral mixing may require a fair distance (about 100 times the bed width). For that reason, water quality in rivers is highly variable in time and space and cannot be adequately represented by mean conditions. Rivers transport pollutants, especially the persistent ones, from the point of pollutant entry to the point of discharge into other rivers, aquifers, wetlands, reservoirs, lakes, estuaries and marine waters. Therefore, upstream users must consider downstream uses.

The natural variability of water quality in a river depends on the combination of diverse environmental factors (Meybeck and Helmer, 1989), including the presence of highly soluble minerals, the thickness of the substrate above the bedrock, the ratio of the annual precipitation volume of the river catchment, the annual river streamflow, the presence of peatbogs, wetlands and marshes, and contributions of nitrogen, phosphorus, silicon, calcium, magnesium, sodium and potassium from soil erosion.

Urban areas affect the water quality in rivers through discharges of wastewaters, and thereby modify temperature, suspended solids, organic matter, turbidity and faecal pollution indicators in rivers, as shown in Table 4.5. Self-purification processes that take place in rivers and regulate soluble compounds in riverine water include flow

Table 4.5 Changes in river characteristics due to sewage discharges

Disturbance	Description
Physical characteristics	Suspended solids and turbidity increase; temperature increases if there are cooling water discharges.
Faecal pollution	Mainly in cities of developing countries, particularly in those with fast growth. Faecal coliform counts can be as high as 10^6 MPN/100 mL.*
Organic matter	Raw wastewaters as well as treated wastewater (to a lesser extent) produce dissolved oxygen demand and nitrogen release in rivers. The effect is directly related to the ratio of the polluting discharge to river discharge. Disturbances can be observed from some kilometres to up to 100 km from the discharge point. Turbulence helps to re-oxygenate the riverine water.
Eutrophication	Between 1950 and 1960, the first reports of eutrophication in lakes and reservoirs were made; later, in the 1970s, the same problem was reported for rivers.
Increase in salinity	Treated sewage and mining and industrial discharges contain salts, which increase the salinity of water supplies, especially in arid and semi-arid areas with high rates of evaporation.

* MPN: most probable number

turbulence, evaporation, absorption to sediments, primary production and organic matter oxidation (Chapman, 1992). Turbulence contributes to volatilization of chemicals and higher dissolved oxygen levels. Evaporation may increase pH, electrical conductivity and precipitation of soluble materials. Sediment absorption can result in reduced concentrations of nutrients, dissolved organic carbon, soluble metals and organic micropollutants. Primary production may increase precipitates, reduce nutrients by consumption, and increase DO and dissolved organic carbon (DOC). Finally, organic matter oxidation in the water column or anoxic sediment reduces pH, increases dissolved nutrients, reduces DO and DOC, and potentially increases soluble metal concentrations by de-sorption (Chapman, 1992).

DISSOLVED OXYGEN

All rivers possess a self-purification capacity for biodegrading organic matter through aerobic oxidation. Thus, this process depends on the concentration of dissolved oxygen and can be represented by the oxygen sag curve shown in Figure 4.4. In this curve, four zones can be distinguished.

- The degradation zone, where the discharge enters the river, is characterized by visible signs of pollution (floating solids), turbidity is high and dissolved oxygen begins to diminish due to the presence of oxygen-consuming bacteria.
- The decomposition zone, where dissolved oxygen is very low (sometimes practically absent), contains no superior aquatic fauna. Water is black, gives off an odour, and a great amount of sediment is deposited on the river-bed.
- The recovery zone, where DO increases, and micro-organisms are present in small numbers (facultative anaerobic or strict aerobic), contains superior forms of biota (small larvae, worms and fish).
- The clean water zone, whose appearance is similar to the river upstream of the discharge point, is characterized by no floating solids, clear water, DO near saturation, superior micro-organisms and other forms of biota. Fish are generally more abundant than upstream of the discharge point.

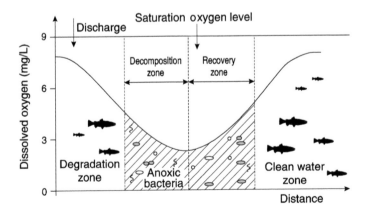

Figure 4.4 Dissolved oxygen sag curve

The form of the sag curve is a function of the pollution load, the quality of the receiving water, temperature and the pollution discharge to river flow ratio. Temperature plays an important role in the sag curve because in warmer water oxygen dissolves in smaller quantities than in cold water, although biological activity and therefore oxygen demand increase. Mixing is another important factor since it promotes oxygen transfer and dilution.

NUTRIENTS

Excess nutrients may cause eutrophication in rivers, particularly where rivers are impounded with retention times greater than one month. Impoundments also increase water evaporation and dramatically modify the composition of aquatic organisms. Oxygen content is reduced, the pH is modified and ammonia is released (NH_3), which are the conditions that have dramatic impacts on fish. In very slow river flows, an excess of phytoplankton caused by eutrophication causes taste and odour problems in drinking water. For a detailed discussion of water quality changes in river impoundments, see Chapman (1992).

Other effects encountered in rivers include increased salinity (total dissolved solids, TDS), acidification, and higher concentrations of trace elements and nitrates. Sewage, and industrial and mining wastewater discharges as well as winter runoff in cold climate countries (where salt is used in winter maintenance) increase Cl^- and Na^+, SO_4^{2-} and CO_3^{2-} content in rivers. Other reasons for increased TDS in rivers may be soil erosion and water evaporation. Where such waters are used for water supply, the costs of treatment to remove TDS increase.

Acidic waters may enter rivers as a result of mining and industrial discharges, and atmospheric deposition. This may change the pH of the riverine water, where the buffer capacity is low, as encountered in areas with non-carbonated detritic rocks (sandstones) or crystalline rocks (granites or gneisses). Acidification of rivers diminishes water biota and reduces its diversity. It also increases metal dissolution, possibly producing toxic effects and limiting uses of such water.

Trace elements are discharged into rivers from various industrial operations, urban runoff, pesticide applications (copper), atmospheric deposition, and sanitary landfill leachates. Most such chemicals (metals and organic compounds) are adsorbed to particles at normal pH and redox potential conditions. Trace metals consequently accumulate in sediments, from which they can be released if pH or redox potential conditions change. Bacteria also can solubilize metals (Hg, As, and Pb) in sediments and transform them into volatile organometallic compounds. Thus, sediments play an important role in water pollution and decontamination.

Finally, in some rivers, nitrates may reach high values as a result of polluted domestic and industrial discharges. Anoxic conditions may be created in rivers and cause nitrogen to be released from river-beds into the water column by denitrification. When oxidized nitrogen reaches 10 mg N/L in river water, it becomes unsafe for human consumption.

4.4.5.2 Lakes and reservoirs

Lakes and reservoirs are characterized by long hydraulic residence times and relatively low capacity for decontamination. Lakes located near urban areas are often used for water supply and recreation, and as receiving waters for urban effluents. Thus, a delicate balance among the various water uses must be achieved. Deep lakes and reservoirs are typically thermally stratified, with the highest density water located at the bottom (hypolimnion), and lower density water found in the upper, warmer layer (epilimnion). The epilimnion is completely mixed by wind and waves; the hypolimnion is cooler than the epilimnion. There is a clear physical separation between the two layers, known as a thermocline, whose thickness typically varies from 10 to 20 m in reservoirs deeper than 35–50 m. In shallow lakes, stratification does not occur because of wind-induced mixing (Chapman, 1992). As temperature changes during the year, many lakes undergo alternating periods of vertical mixing and stratification. As mentioned in Section 4.4.1.4, stratification can be also caused by dissolved solids (particularly chloride), which impede vertical mixing. One of the main characteristics of lakes and reservoirs is significant water loss through evaporation.

The main water quality problems in encountered in lakes and reservoirs are summarized in Table 4.6.

Acidification is one of the most important problems observed in lakes in temperate climates. It results from acidic depositions caused by air pollution and occurs in lakes with low alkalinity, hardness, conductivity and dissolved solids. Normally, these conditions are found in lakes located in zones with noncarbonated soils (crystalline rocks or quartz sandstones) in temperate climates.

An increase in the salinity of lake waters is a natural process that occurs to a greater extent in rivers and lakes with high evaporation rates. Saline inflows into lakes accelerate this process and limit possible water uses.

Table 4.6 **Main water quality problems in lakes and reservoirs**

Process	Cause	Effects on water quality
Acidification	Atmospheric deposition of acidity.	Decrease in pH, increased concentrations of heavy metals, loss of biota.
Increased salinity	Water balance modifications, soil leaching, industrial and municipal discharges.	Increased TDS, increased treatment cost if dissolved solids exceed 1,500–2,000 mg/L.
Eutrophication	Excess of nutrients.	Increased algae and plant production; consumption of the hypolimnion oxygen; release of Fe, Mn, NH_4 and metals from the hypolimnion; loss of diversity in the higher trophic levels; favourable conditions for reproduction of mosquitoes and spread of shistosomiasis.
Pathogen contamination	Sewage discharges.	Spread of disease including infection by bacteria, virus, protozoa and helminth ova.
Increased toxicity	Industrial and municipal discharges.	Increased concentrations of trace organic and metal toxicants; introduction of endocrine disruptors; toxic effects through bioaccumulation and biomagnification; fish tumours and loss of biota.

Eutrophication is a natural process of water body 'aging' and can be accelerated by contamination. Eutrophying lakes pass through four stages: oligotrophic, mesotrophic, eutrophic and hypereutrophic. Oligotrophic lakes are characterized by low nutrient concentrations and little biological productivity. When the nutrient content is increased by soil leaching and polluted inflows, flora and fauna also increase in lakes. If large amounts of nutrients are added, especially nitrogen and phosphorus, algae and aquatic plants begin to proliferate abnormally, covering the lake surface and preventing sunlight and oxygen from entering the water. Normal life in lakes is then altered. Under these conditions, during the day, primary production exceeds the detritus bacterial decomposition of algae. Oxygen content can reach 200% of saturation and pH values of 10 or more in the early afternoon. At night this situation is reversed; oxygen drops to 50% of saturation and pH diminishes to below 8.5. If in addition the lake also receives biodegradable organic contaminants, it is difficult to achieve oxygen saturation during the day. The flora die due to the lack of sunlight and sediments turn anaerobic due to decomposition. The normal life of lakes is disturbed and the lake is said to be eutrophied. After some time, sediment accumulation and the high rate of evapotranspiration caused by surface plants dry up the lake and transform it into a marsh. Other factors related to the development of aquatic plants during eutrophication are discussed in Section 4.8.3.2. Common lake changes caused by eutrophication include:

- loss of biodiversity and displacement of native species
- obstruction/blockage of channels and drains in irrigation zones and feed channels of hydroelectric plants, when connected to eutrophic lakes
- restrictions on tourist, recreational and fishing activities
- reduction in the useful life of the lake.

Reservoirs behave similarly to lakes. Constructed reservoirs require some aging to develop the characteristics of natural lakes. As these reservoirs usually serve many purposes, including water supply, flood control and hydroelectric power generation, their levels fluctuate greatly and frequently. Water withdrawals alter the natural thermal stratification. Perhaps the most significant effect of damming a river is the transformation from flowing to stagnant water, with a concomitant reduction in self-purification capacity and promotion of eutrophication. More detailed description of the effects of dam construction on rivers can be found in Chapman (1992).

4.5 URBANIZATION EFFECTS ON WETLANDS

Wetlands represent 6% of the terrestrial surface on Earth and occur in all types of climates, from the tropics to the tundra. The term 'wetland' describes a great variety of specific ecosystems such as bogs, bottomlands, fens, floodplains, mangroves, marshes, mires, moors, muskegs, beaches, peatlands, pocosins, potholes, reedswamps, sloughs, wet meadows and wet prairies (Mays, 1996). Wetland ecosystems represent the transition between terrestrial and aquatic systems and are inundated, or characterized by high water tables (at or near the land surface) during much of the year. The general importance of wetlands can be inferred from the list of their functions and values presented in Table 4.7.

Table 4.7 Attributes, functions and values of wetland ecosystems

Wetland functions

1. *Hydrological flux and storage*
   a. Aquifer recharge by wetlands, conveyance of flows.
   b. Water storage and flow regulation.
   c. Regional stream hydrology (discharge and recharge control).
   d. Regional climate control (evapotranspiration export, large-scale atmospheric releases of H_2).

2. *Biological productivity*
   a. Net primary productivity.
   b. Carbon storage.
   c. Carbon fixation.
   d. Secondary productivity.

3. *Biogeochemical cycling and storage*
   a. Nutrient source or sink in landscape.
   b. C, N, S, P etc. transformations (oxidation/reduction reactions).
   c. Denitrification.
   d. Sediments and organic matter storage.

4. *Decomposition*
   a. Carbon release (global climate impacts).
   b. Detritus output for aquatic organisms (downstream energy source).
   c. Mineralization and release of N, S, C, etc.

5. *Community/wildlife habitat*
   a. Habitat for critical species (unique and endangered).
   b. Habitat for algae, bacteria, fungi, fish, shellfish, wildlife and wetland plants.
   c. Biodiversity enhancement.

Wetland values

1. Flood control, conveyance and storage (1, 2).*
2. Sediment control (filter wastes) (3, 2).
3. Wastewater treatment system (3, 2).
4. Nutrient removal from agricultural runoff and wastewaters (3, 2).
5. Recreation (5, 1).
6. Open space (1, 2, 5).
7. Visual/cultural amenity (1, 5).
8. Hunting (water fowl, beaver, muskrats) (5, 2).
9. Preservation of flora and fauna (endemic, refuge) (5).
10. Timber production (2, 1).
11. Shrub crops (cranberry and blueberry) (2, 1).
12. Production of medical supplies (streptomycin) (5, 4).
13. Education and research (1–5).
14. Erosion control (1, 2, 3).
15. Food production (shrimp, fish, ducks) (2, 5).
16. Archaeological, historical and cultural resources (2).
17. Habitat of threatened, rare, and endangered species (5).
18. Water quality enhancement (3, 1, 4).
19. Water supply (1).
20. Global carbon storage (4, 2).

*Wetland values are directly related to wetland functions (1–5) or those functions that can be adversely affected by the overexploitation of values. The function numbers are ranked in descending order of importance.

Source: Richardson and McCarthy, 1994.

Plants and animals that inhabit wetlands are uniquely adapted to live under conditions of intermittent flooding, the lack of oxygen (anoxia), and harsh (often toxic) conditions characterized by presence of reduced species of chemicals (sulphurs rather than sulphates). These characteristics make wetlands suitable for treating wastewater; constructed wetlands have been used for enhancing stormwater quality, and treating CSOs and wastewater. Constructed wetlands are designed to mimic the characteristics of natural ones, but are typically strongly affected by accumulation of solids and chemicals resulting from the treatment of urban effluents. Such wetlands, with either surface or subsurface flow, have been recommended for stormwater quality enhancement, and good-guidance is available for their design (MOE, 2003), and their effectiveness in removal of suspended solids and some other constituents is well documented (US NURP) (US EPA, 1983). Concerns have been raised about the quality of habitat in constructed wetlands when they become colonized by wildlife, in view of the accumulation of polluted sediment, chloride and other pollutants (Bishop et al., 2000a, 2000b).

One of the key functions of wetlands is their capacity to store, transform and recycle nutrients. In addition, they provide effective treatment of sewage and urban runoff. Wetlands retain 60–90% of incoming suspended solids and 80% of eroded runoff materials, and adsorb heavy metals (Mays, 1996), and also remove pathogens (Table 4.8). However, the treatment efficiency of wetlands varies greatly, depending on climatic conditions, the season of the year, the type of wetland and the hydraulic residence time. Furthermore, when treatment capacity is exceeded, odour is produced. Wetlands may also encourage the breeding of mosquitoes, which sometimes serve as disease vectors (e.g. West Nile virus) and require implementation of mosquito control programmes.

Other reported applications of wetlands include the treatment of municipal wastewater, septage, acid mine drainage, ash pond seepage and pulp mill effluents (Metcalf and Eddy, 2003). To operate such treatment systems effectively, implementation of pre-treatment and good maintenance are essential. Maintenance operations include removal and disposal of deposited sediment with associated contaminants, occasional harvesting of vegetation, clearing of drainage channels and similar measures.

Ecological problems encountered in constructed wetlands are similar to those found in natural wetlands receiving polluted urban effluents; accumulation of pollutants from such sources will affect the operation and integrity of wetland systems.

Table 4.8 **Removal of pathogens in wetlands**

Micro-organism	Removal (%)	Reference
Faecal coliforms	90–99	Perfler and Haberl, 1993; Hiley, 1995; Rivera et al., 1995; Karpiscak et al., 1996; Rivera et al., 1997
MS-2 Coliphages	67–84	Gersburg et al., 1989
Cryptosporidium	53–87	Karpiscak et al., 1996
Giardia	58–98	Karpiscak et al., 1996
Acanthamoeba astronyxis, *A. polyphaga* and *A. rhysodes*	60–100	Ramirez et al., 1993
Entamoeba coli, E. histolytica, *E. nana* and *Iodamoeba butschlii*	100	Rivera et al., 1995
Ascaris lumbricoides eggs	100	Rivera et al., 1995

4.6 URBANIZATION EFFECTS ON SOILS

The interactions between soils and various components of the urban hydrological cycle occur in urban areas through such processes as soil erosion, leaching from contaminated soils, water infiltration/percolation and on-land sludge (biosolids) disposal.

4.6.1 Erosion

Two types of soil erosion occur in urbanizing catchments: stream bed erosion described earlier in Section 4.4.1.2 and sheet erosion (detachment of soil particles by rain drops), which occurs mostly on large tracts of soils stripped of vegetation cover and subject to intense soil erosion. Soil losses due to erosion can be described by the Universal Soil Loss Equation (USLE) developed by Wishmeier and Smith (1965) in the following form:

$$q = R \, K \, L \, S \, C$$

where q is the erosion rate (tonnes/ha) from a site, R is the rainfall erosivity (tonnes/ha), also called the rainfall energy factor, K is the soil erodibility (dimensionless) (Wishmeier et al., 1971), L (m) and S (per cent) are the length and slope of the site, and C is an erosion control factor (dimensionless).

Eroded soil is transported by runoff into storm or combined sewers, and in the second case, to the sewage treatment plant. Thus, erosion leads to a loss of a resource (topsoil), and at the same time causes operational problems in sewers and sewage treatment plants. To mitigate these problems, excessive soil erosion needs to be prevented by implementing erosion and sediment control programmes in all construction activities. Where sediment deposition in sewers impairs the system hydraulic capacity, sediments may have to be removed by mechanical or hydraulic means, often at significant costs. Furthermore, the removed sediment may be contaminated and may require special conditions for safe disposal. Some of these issues are documented by an example of sewer sediment problems in Mexico City, where about 40% of sewer sediments originate from soil erosion and the rest is wastewater sludge.

SEDIMENTS REMOVED FROM THE MEXICO CITY SEWER SYSTEM

A recent study (Jiménez et al., 2004b) of sediments removed from the Mexico City sewer system reveals the magnitude of the problem that erosion, solid wastes, and water and wastewater sludge from treatment plants can represent for a city. The city has about 19 million people and is served by a combined sewer system discharging annually $1,700 \times 10^6$ m^3 of sewage, of which 70% is municipal wastewater and 26% is stormwater. The annual precipitation is about 750 mm/y. The sewer systems must be cleaned each year to prevent flooding in the city. About 2.8×10^6 m^3 of sediments are generated annually, but only 30% is extracted because of budget limitations. These sediments originate in discharges of sludge from twenty-seven wastewater treatment plants, soil erosion, uncollected municipal solid wastes and sedimentation of suspended solids contained in wastewater. It has been estimated that at least 40% of these sediments are contributed by erosion soil losses in natural conservation areas in the city. About 22% of the city territory is subject to erosion soil losses at a rate of 500–3,500 t/ha/year. The removed sediments are disposed of at the only available landfill near the city, which has a limited capacity. The capacity for dealing with this problem is only now being developed. The composition of removed sediments varies greatly; some are too contaminated for normal treatment and require special disposal; others can be used in landfills compensating for land subsidence, or for soil conditioning, landfill cover, or in nurseries.

4.6.2 Transport of pollutants in soils

Numerous sources of soil pollutants can be found in urban areas: industries, storage tanks, petrochemical facilities, landfills, natural resource extraction, garbage dumps, hazardous waste confinement sites, factories, oil transport pipes and small businesses. Locating polluted soils in cities is a complex problem, but determining the physical, chemical and biological aspects of such contamination is even more difficult.

Soil contamination begins in the unsaturated zone, where the pollutants may exist in four phases (Norris et al., 1994): (a) as a gas in the soil pores, (b) adsorbed or adhered to solid surfaces, (c) dissolved in water (hydrosoluble compounds), and (d) dissolved in other liquids (hydrophobic compounds). Pollutant transport occurs in the gaseous and liquid phases (water and liquids); from the gaseous phase, pollutants are transported by volatilization to the atmosphere, while those in liquid phases move deeper into soils by diverse mechanisms. The magnitude of their infiltration and the mechanism used depends on the polluting agent and soil properties. The driving forces for the polluting agent are its solubility, vapour pressure density, viscosity, persistence and hydrophobicity. For soils, geology, mineralogical composition, distance to the aquifer, organic matter content and hydrology are the intervening factors.

Liquids that are lighter than water and hydrophobic, like hydrocarbons, first spread laterally to form a layer that floats over the phreatic level. From there, vertical migration occurs until capillarity forces retain the pollutant, or the insoluble fraction is exhausted. Phreatic level fluctuations caused by seasonal variations, pumping or intentional or non-intentional recharge enhance the transport and mixing of pollutants. Vertical transport continues through the unsaturated soil zone by capillary and gravity forces until the hydrophobic compound arrives at the capillarity strip or at a zone of low permeability. There, the compound forms a layer that extends and thickens until a sufficient load is reached to penetrate through the capillarity strip to the phreatic level. As the density of many hydrophobic solvents (like chlorinated ones) is greater than that of water, the polluting agent continues sinking until it reaches an impermeable layer. In this place it forms deposits of the polluting agents that are difficult to detect and remove.

Soluble compounds, including metal compounds, also migrate vertically but along with water and/or by diffusion. Since metals tend to adhere to soil particles, their transport with water is possible only in an acidic environment, and in soils containing low concentrations of organic matter and calcium. Because the water-soluble compounds are well distributed in the groundwater, they do not accumulate in particular places like the hydrophobic compounds. For that reason, remediation of soluble compounds requires pumping and treating groundwater, while in the case of hydrophobic compounds, remediation implies soil extraction or *in situ* treatment (or remediation).

For all chemical compounds there is a direct relation between soil permeability and the potential for groundwater to be polluted. For that reason, sandy and gravelly soils are more sensitive to pollution than clay soils. Furthermore, chemical transport depends on other phenomena that delay or accelerate the chemical movement, including photochemical reactions, biological assimilation, oxidation–reduction reactions, ion exchange, complexation, precipitation, and aerobic and anaerobic degradation.

4.6.3 Changes in water quality during percolation through soils

While some soils may have the potential to pollute water, soils can also have the opposite effect, because they may enhance the quality of infiltrating water. They may remove suspended solids, nutrients, metals, organic compounds and pathogens from the percolating water through filtration, biological degradation, ion exchange and pathogen dieoff. These processes are sometimes used as a wastewater treatment method also known as soil aquifer treatment (SAT). Treatment is achieved by combining inundation and dry cycles to maintain aerobic conditions in soils if the soil percolation rate and hydraulic and mass loads in the influent are limited. To avoid pollution of the aquifer, a certain depth to the saturated zone must also be maintained.

Reuse of treated or untreated wastewater for agricultural irrigations is in fact a low-cost treatment alternative (Bouwer et al., 1980). In developed countries, this treatment process is controlled; however, in developing countries, it is mostly not. Wastewater treatment is achieved by the great amount of carbon and nitrogen consumption that the terrestrial ecosystem exerts (Bouwer and Rice, 1984), significantly in excess of that in a water body. It should be noted that even in unfavourable conditions in certain types of soils, there is always some improvement in the quality of water percolating through such soils (Oron, 2001). Soils remove more than 90% of suspended solids, 80% of organic carbon, 70% of total nitrogen, and almost 50% of phosphorus from percolating wastewater (Bouwer and Rice, 1984). The passage of treated wastewater through the vadose zone (at least 3 m thick) removes persistent organic residuals and pathogens (Wilson et al., 1995; Fujita et al., 1996; Quanrud et al., 1996). Due to the high infiltration rates (30 to 110 m/year) used in some types of SAT systems most of the infiltrating water reaches the aquifer and therefore the application of effluents on soils may be viewed as a recharge method that increases the groundwater availability (Lance and Gerba, 1980). Also, SAT systems were found to be efficient in treating and reclaiming urban stormwater (Oron, 2001).

Two aspects must be watched during effluent infiltration: organic recalcitrants and viruses. Viruses can migrate long distances in soils; vertical distances of up to 67 m and horizontal distances of 408 m have been reported. Table 4.9 describes some of the phenomena that occur during virus transport in soils.

4.6.4 Effects of sludge disposal on soils

Wastewater treatment produces treated water and a solid residual, which is called sludge or biosolids. The latter term refers to well-treated and processed sludge that can be recycled, for example by disposal on agricultural land. Sludge may contain all the constituents removed from wastewater as well as those added during treatment. In many countries, sludge is poorly managed and often disposed of, without any treatment, onto soils or into water bodies. In other cases, sludge is disposed of in lagoons,

Table 4.9 **Processes or characteristics affecting virus transport in soils and groundwater**

Process or characteristic	Effect
Adsorption	Increases virus survival and slows down its transport; depends on ionic forces, pH, humidity and salinity.
Virus aggregation	Makes viruses more resistant.
Temperature	Represents the determining factor in inactivating viruses, in comparison to the effects of pH, sulphates, iron, hardness and dissolved solid content.
Microbial activity	Eliminates viruses.
Soil moisture	Absence of water inactivates viruses; saturated soils support virus transport.
pH	Affects adhesion and activity properties, depending on the type of virus.
Dissolved salts	Iron and aluminium salts inactivate viruses and slow down transport.
Organic matter	Prevents absorption.
Hydraulic conditions	Transport is increased with flow.
Types of viruses	Each species behaves and is transported differently.
Soil properties	In soils with coarse texture, there is greater mobility than in karstic fine-textured soils; clay retains viruses.

Sources: Drewry and Eliassen, 1968; Nestor and Costin, 1971; Berg, 1973; Bixby and O'Brien, 1979; Gerba and Bitton, 1984; Yates et al., 1985.

landfills or at uncontrolled sites. Better-quality sludge is used to control land degradation or improve agricultural soil characteristics. Experience shows that more hazardous practices are more common in developing countries than in developed ones.

4.6.4.1 Sludge production

Reliable information on sludge production is available more or less only for the United States and the European Union. In 1998, sludge production was estimated in the United States as 7 million dry tonnes (US EPA, 1999), which is comparable to the total quantity generated in the European Union in 1992, 7.4 million dry tonnes.

4.6.4.2 Sludge quality

The environmental impact of sludge disposal on soils depends on its composition, type of treatment and disposal method. The chemical composition of sludge is not well known, due to its variability and the lack of research. It depends more on the sludge's origin than on the wastewater treatment process. Chemical compounds occur in sludge as precipitates (sulphides, oxides or bicarbonates) or are adsorbed to, or chelated with, organic matter. Six classes of chemical compounds are distinguished: (a) metals and cyanides, (b) volatile organic compounds, (c) semi-volatile organic compounds, (d) pesticides and PCBs, (e) organic matter and nutrients, and (f) others. In general, in developing countries, toxic compounds are present in lower quantities than in developed countries, unless there is a strong input of specific industrial wastewaters (such as, for instance, tannery wastewaters). Concerning heavy metals, the limited data available from Brazil, Chile, China, Mexico and South Africa indicate trace metal concentrations well below the international standards, without any treatment (Jiménez et al., 2004a).

Sludge contains a lot of micro-organisms. Pathogen content is the main concern since sludge is often spread through direct hand-contact in the developing world, where health protection is relatively low. The difference between microbial counts in sludge in developed and developing countries is very important. Orders of magnitude vary from 10^5–10^6 to 10^7–10^{10} MPN (most probable number)/g of solids for faecal coliforms, 10^3–10^7 MPN/g of solids for *Salmonella typhi*, 10^2–10^4 cysts/g of solids for *Giardia lamblia* cysts, and from <1 to 177 ova/g of solids for viable helminth ova, respectively (Jiménez et al., 2004a).

4.6.4.3 Biosolids (sludge) application on land

Biosolids (well-treated sewage sludge) can be reused for diverse purposes, including applications on agricultural land, or production of construction materials (e.g. bricks). Applications on land include agricultural, forestry and soil restoration uses. Table 4.10 shows the rates at which biosolids are applied and compares them with the rate of confinement in landfills.

Biosolids are applied on (a) sites with no public access, like agricultural fields (land grass and culture soils), forests, restoration sites (mining and construction sites), airports and highway corridors; and (b) sites with public access, such as public parks, greenhouses, golf courses, prairies, gardens and cemeteries.

Applications of biosolids in agriculture are very important for reducing the cost of fertilization, controlling soil degradation by erosion, improving soil characteristics, and increasing agricultural productivity. At the same time, there are some risks involved due to concentrations of heavy metals and trace organics, and disease vector attraction. These are all important issues, which however are outside of the scope of this report. More information on this subject can be found in Girovich (1996).

4.6.4.4 Sludge disposal

Other methods of sludge disposal include landfills, lagooning, incineration and dumping into the sea. The last two options are prohibited in some jurisdictions, but still used in others due to the lack of disposal space. Land disposal can be designed as a monofill

Table 4.10 **Typical rates of biosolids applications**

Final disposal method	Frequency of application or final disposal	Disposal rate (t/ha) Range	Typical
Agricultural use	Annual	2–60	10
Forests	One, three or five times per year	8–200	40
Soil remediation or restoration	Once	6–400	100
Confinement in landfills	Annual	200–800	300

Source: US EPA, 1997.

(sludge only), or sludge can be sent to a local landfill where it competes for space with solid wastes. Ponds and lagoons are also sometimes used as disposal sites. The lack of space for land disposal is an increasing problem in some cities. Since many of them are no longer permitted to use local landfills, they have to look for other disposal alternatives (Lue-Hing et al., 1992). Hogg (2002) reported the costs of sludge handling in the range from \$105 to \$350/tonne, for a semi-solid sludge application on land and monoincineration, respectively. Further increases of costs of sludge disposal are expected in the future.

4.6.4.5 New chemicals of concern in sludge

Municipal sludge contains various toxic chemicals, which may cause a variety of environmental effects. The discussion of such effects is beyond the scope of this report, but some emerging challenges are briefly listed: endocrine disrupting substances (EDSs), pharmaceuticals (including antibiotics), and personal care products.

EDSs are compounds that interfere with animal hormonal activity and are present in municipal, industrial and sewage discharges (Chambers et al., 1997). They may damage reproductive organs and alter reproductive functions. In human beings, the main concerns about endocrine disrupters are their similarity to feminine and masculine hormones (estrogens and androgens); however, little is known about their actual effects. Endocrine disrupters are difficult to detect because they appear in extremely low concentrations in water (in the order of ppb), and, the traditional parameters used to measure pollution, including toxicity tests, do not provide useful information as to their presence. Yet even very low concentrations of EDSs (in ppb) may be sufficient to cause environmental effects (Chambers et al., 1997). It is currently known that many xenobiotic compounds, such as DDTs, DDEs and chlorinated pesticides, certain pharmaceuticals and musk compounds are also endocrine disruptors because they mimic estrogen. Also, some natural compounds, such as 17-β estradiol and estrone are hormones produced by micro-organisms and plants. Matsui et al. (2000) measured the content of compounds similar to estrogen in raw and treated wastewater, reporting values of 50–150 ng/L before and 1–13 ng/L after treatment.

Residues of pharmaceuticals and personal care products (PPCPs) represent another emerging environmental issue. Extensive studies (Ternes and Joss, 2006) have established the presence of these chemicals in water resources and their main pathway of entry into the environment via municipal wastewater. Conventional wastewater treatment processes with nutrient removal remove at least part of the PPCPs, mostly by biologic transformation and sorption onto sludge. Consequently, sludge retention of at least ten days (required for nitrification) is required. Advanced treatment processes, including ozonation, activated carbon filtrations, and tight membrane filtration, represent potential solutions to the need to attenuate PPCPs in treated effluents (Ternes and Joss, 2006). However, before deciding on risk control measures, more information is needed on environmental risk assessment for PPCPs. Thus, further research of these new chemicals of concern, their effects, and means of mitigation is needed.

4.7 URBAN IMPACTS ON GROUNDWATER

Two types of urban effluent discharges reach the groundwater: non-intentional (accidental) and intentional. Non-intentional discharges are much more significant than expected. Typical examples of such discharges include infiltration of liquids from storage tanks and lagoons, sewage reuse in agriculture or landscape irrigation, landfill leaching, and infiltration of polluted water from channels and rivers and cemeteries.

4.7.1 Unintentional discharges into groundwater aquifers

Foster et al. (2002) concluded that:

- A large number of urban human activities may potentially pollute aquifers through non-intentional recharge, but only a few are responsible for the most severe problems.
- The severity of the pollution caused is not directly proportional to the source size, because extended discharges from small operations (e.g. machine shops) can cause great impacts.
- The larger and more complex industries pose lower risks of accidental or intentional spills, because of monitoring and quality control programmes.
- The concentration of pollutants in groundwater depends on the pollutant dispersion and persistence more than on the type of pollution source.

Fortunately, the soil vadose zone retains a large part of the pollutants, depending on the type of chemical compound (concentration, volatility and density), geological properties of the soil material (physical heterogeneity and hydraulic characteristics), reactions involved (sorption, ion exchange, precipitation, oxidation, reduction and biological transformation), and the hydrological conditions (retention time, flow pattern and evaporation). Water quality modifications caused by passage through soils were discussed in more detail in Section 4.6.3, whereas the following section focuses on sources of aquifer pollution.

Any landfill, controlled or not, can be a source of aquifer contamination. Most such contamination comes from old landfills or garbage dumps which were built without adequate guidelines and before hazardous waste segregation was introduced. Particularly in developing countries, landfills built from 1950 to 1970 are not equipped with geomembranes and are sources of leachates. Such problems are worse in humid regions, where rain generates larger amounts of leachates. The situation is not much better in arid climates, where leachates may be smaller in volume but may possess higher concentrations of contaminants. The potential impacts of sanitary landfills on aquifers can be estimated from the type of solid wastes deposited and rainfall amounts (Nicholson et al., 1983).

Tanks and lagoons are used to treat, evaporate or store liquid wastes (municipal, industrial and mining drainage) or store water for flood control, particularly in developing countries. Generally, such storage facilities are less than 5 m deep and their hydraulic residence times vary from 1 to 100 days. Even if some of these structures are initially waterproof, almost all develop some leakage, the magnitude of which depends on the type and quality of construction and the quality of maintenance. The typical volume of leakage has been estimated at 10 to 20 mm/d by various authors (Miller and Scalf, 1974; Geake and Foster, 1986).

Other sources of groundwater contaminants are confinement sites of hazardous wastes. Various trace metals and organics may leak from these sites and enter aquifers (NRC, 1994).

Given the rapid population growth and the lack of sewers in urban areas in most developing countries, wastewater may leak from septic tanks, latrines, sewers and wells, or be directly discharged into soils (Lewis et al., 1986). Such leaks contribute to aquifer recharge and pollution. For example, Capella (2002) estimated that sewage leakage into the Mexico City aquifer amounts to $1 m^3/s$ and Foster (2001) estimated that, in extreme cases, such leakage can be equivalent to 500 mm of rain/year in highly populated areas of developed nations. Thus, in areas with limited or non-existent drainage and population densities higher than 100 persons/ha, there is a high risk of groundwater pollution. This risk diminishes and is of local significance only in predominantly residential areas with high-density drainage (Foster, 2001).

Pollutants commonly associated with sewage leakage (exfiltration) include biodegradable organic matter, nitrogen compounds, phosphorus, micro-organisms (including those causing typhoid, tuberculosis, cholera and hepatitis), suspended solids and trace organic compounds. Among them, nitrates are the most mobile and persistent, which is why they are normally detected in polluted aquifers.

Underground tanks are used for storing various liquids, but most often gasoline. Usually, leaks develop due to corrosion and poor connections, and there is a close correlation between the frequency and size of the leaks and the age the tank (Kostecki and Calabrese, 1989; Cheremisinoff, 1992). In general, tanks that are more than twenty years old are very likely to leak, especially if not appropriately maintained. Leakage problems can be significantly reduced with better design, construction, operation and maintenance. Tank leaks can be controlled using cathodic protection or double-wall steel or plastic reinforced with fibreglass.

Fuel station tanks cause many cases of groundwater contamination (Fetter, 1988). For example, in the United States, they account for at least one out of thirty leaks (Bedient et al., 1994). This pollution problem is exacerbated by the fact that gasoline tanks are widely distributed in cities, reflecting the local fuel demand rather than the need for environmentally suitable locations. Even though these leaks are usually small, they occur over long periods of time and produce pollution plumes of great extent. To avoid such pollution, leaks should be detected by standard procedures and the tanks should be sealed. Additional risks occur when pumping stations are coupled with service stations, where significant amounts of organic solvents may be spilled on soils.

The conveyance of contaminated water in open channels and rivers is frequently a source of aquifer pollution and recharge. The significance of such an impact depends on the volume and quality of the seeping flow, and can be determined only by monitoring each case.

Industrial activities can seriously pollute subsoil and groundwater, depending on the type, volume and the way of handling liquid and solid wastes. Such risks are particularly significant in industries using more than 100 kg of toxics per day (hydrocarbons, organic synthetic solvents, heavy metals, etc). The pollutants involved are related to the type of the industry, as discussed by Foster et al. (2002).

In many cities and surrounding urban areas, small industries and service operations (mechanical factories, dry cleaning services, etc.) handle toxic substances like chlorinated solvents, aromatic hydrocarbons and pesticides. It is important to control their wastes (liquid and solids), since they frequently store or discharge them into soils instead of recycling or disposal at appropriate confinement sites. Nevertheless, it is difficult to control small industries and services, because they often move, relocate or operate intermittently. Without proper regulations, it may be difficult to exert such controls.

The burial sites of humans and animals represent sources of microbiological contamination of aquifers, although to a small extent. To avoid this problem, watertight caskets must be used, but they may not be affordable in poor countries.

4.7.2 Intentional discharges into groundwater aquifers

In urban areas where there is intensive soil use, aquifers may be used as water storage facilities, also called aquifer storage recovery systems (ASRs). This is possible because any aquifer is essentially a water reservoir, from which water can be withdrawn when required. Stored water can be of diverse quality, depending on its intended use (Table 4.11). The first systems of this nature began operating in the United States in 1968. Some of the water stored is saline and considered of low quality due to the presence (in some cases of natural origin) of nitrates, barium, hydrogen sulphide, iron and manganese (Pyne, 1995).

For designing ASRs, a well-defined methodology has been established and addresses design, construction and operation. Water quality improvements during water injection depend on the combination of soil filtration properties and the method of water recharge (Bouwer, 1989). Aquifers can be recharged by infiltration through soils or direct injection. The latter requires water of better quality than that applied through soil infiltration, to avoid well blockage and entry of pollutants into the subsoil and aquifer. The injected water must match the drinking water standards, or at least be of the same quality as that of the aquifer (Crook et al., 1995).

Some critical considerations of ASR applications in conjunction with reclaimed water include (Oron, 2001):

- the risk of introducing recalcitrant pollutants into aquifers and soils, if there is no pre-treatment of non-residential discharges to the drainage system
- the leaching of dangerous pollutants originating in households
- the dilemma of disinfecting the effluent prior to recharge knowing that this kills native soil micro-organisms useful in water treatment.

Table 4.11 Objectives of aquifer storage recovery systems (ASRs)

Objectives	
Temporary water storage during various seasons of the year	Control of nutrient leaching
	Enhancing water well production
Long-term storage	Retarding water supply system expansion
Storage for emergencies or as strategic reserves	Storing reclaimed water
Daily storage	Soil treatment
Reduction of disinfection by-products	Refining water quality
Restoration of phreatic levels	Stabilising aggressive waters
Pressure and flow maintenance in the distribution network	Hydraulic control of pollution plumes
	Water temperature maintenance to support fisheries
Improvement of the water quality	
Prevention of saline intrusion	Reducing environmental effects caused by spills
Water supply for agriculture	Compensating for salinity leaching from soils

Source: adapted from Bouwer, 1989; Pyne, 1995; Oron, 2001.

4.7.3 Impacts on aquifers

In many countries, groundwater represents the main source of water supply (Table 3.3). In spite of this fact, many aquifers around the world are currently being over-exploited and polluted. Aquifers are characterized by stable flow patterns, in terms of flow speed and direction. Velocities often vary from 10^{-10} to 10^{-3} m/s and depend to a great extent on the soil porosity and permeability. As a consequence, mixing is impaired in soils. Bedient et al. (1994) listed thirty potential sources of aquifer pollutants, divided into six categories. Among these, the most frequent pollution sources in cities were storage tanks, septic tanks, leaks from sewerage systems, hazardous and municipal landfills, and polluted soils used to store materials.

Based on a literature review, van Eyck et al. (2001) established that potentially harmful contaminants of groundwater include metals, pharmaceuticals, estrogens (natural and synthetic), surfactants, solvents and musks. This selection emphasizes the risk posed by endocrine disrupters, such as nonylphenol (an intermediate product in nonylphenol ethoxylate surfactant biodegradation) and 17-β estradiol (natural human oestrogen), as well as polychlorinated dibenzo-p-dioxin and polychlorinated dibenzofuran, which can be formed by thermal desorption of chlorinated organic compounds.

4.8 URBAN IMPACTS ON BIOTA: LOSS OF BIODIVERSITY

Many urban areas have developed on shores of lakes and rivers and have affected the pre-development biota and entire ecosystems. Also, the UWC may disturb nearby or distant aquatic environments, and such disturbances may cause loss of species or changes in their enzymatic or metabolic systems that alter their way of living. Each organism responds in a different way to the environmental modifications, which is the reason for using biological indicators, including biological communities, to assess the state of aquatic ecosystems.

4.8.1 General structure of water bodies and their biota

Each water body has its own biocenosis, defined as a set of organisms, plants and animals living in harmony, within a specific biotope described by physical and chemical characteristics of the environment. Aquatic ecosystems are physically made up of three zones: the water body including its bottom (aquatic zone), the transition zone between water and earth (amphibious zone), and the terrestrial zone. Each zone is characterized by specific conditions that determine the structure of biological communities.

In water bodies, photosynthetic organisms that depend on dissolved nutrients and solar light to produce organic matter are called primary producers, and represent a food source for zooplankton and small fish that constitute secondary producers, which in turn represent a food source for superior animals like fish, which are called tertiary producers. When all these organisms die, predators are responsible for their degradation and decomposition. All these organisms make up a nutritional chain.

4.8.2 Properties of the water bodies affecting flora and fauna

Organisms require a certain amount of time to develop and reproduce. In moving water with short hydraulic residence times, there may not be enough time for reproduction and organisms may disappear. For normal development, the biota needs dissolved oxygen. Where DO levels are insufficient, even intermittently and for short time periods, there is a rapid decrease in aquatic communities, particularly fish. Dissolution of oxygen in rivers depends on the water movement, temperature and oxygen consumption by living organisms. Organic matter, N and P, are the main sources of nutrients for plankton and benthos in water bodies. Such nutrients must be available in the right amounts to ensure the normal and balanced development of aquatic biota. Finally, toxics are compounds that alter the abundance, diversity and function of organisms. In general, aquatic organisms' response to toxics is slower than to lack of oxygen or sunlight, because they need to undergo sufficient exposure before showing any effects. Nevertheless, the response can be magnified by the presence of other compounds which favour absorption through nutritional chains. Since all the water body characteristics discussed in this section (hydraulic residence time, DO, N and P, and toxics) are changed by urbanization, the flora and fauna in such affected streams will change as well.

4.8.3 Effects of alterations of urban water bodies on biota

The UWC leads to alterations in urban water bodies, which in turn alter aquatic biota. Such changes are discussed here for two types of water bodies: rivers and lakes.

4.8.3.1 Rivers

The biotope and the biocenosis of a river may vary along its course as it moves from the headwaters to the point of its discharge. These variations depend on a number of factors, including climatic and geological conditions as well as the season of the year. In rivers, the main biological community is fish. This community has successfully adapted to colonize flowing water by possessing the following characteristics: (a) growth and reproduction patterns compatible with short hydraulic residence times, (b) use of spaces that serve as a refuge, and (c) the ability to swim against the current. However, fish are very sensitive to physicochemical conditions in rivers, and when these are modified fish are directly or indirectly affected.

Urbanization affects river flow by withdrawals of water and discharges of urban effluents. Flow velocity has direct and indirect influences on biota. If velocity changes are permanent, a new adapted ecosystem evolves and may differ from the original one, depending on the magnitude of the alterations. If the modification consists of withdrawing water, then a minimum flow called the 'ecological flow' needs to be respected; below this level, no development of the native flora and fauna can exist. Rivers with increasing flows also affect amphibious zones, where populations of reeds often grow. High flows can cause erosion and make rivers so deep that the inhabitants of the amphibious zones cannot develop. Also, when the water level in the river rises, the terrestrial zone is affected.

Rivers are also susceptible to eutrophication like lakes, but to a lesser extent because their water is moving. Table 4.12 lists the main effects of eutrophication on rivers.

River-bed rugosity as well as river flow patterns determine the habitat for living organisms. The river-bed undergoes constant changes resulting from erosion and deposition of material by sedimentation or precipitation. These changes are magnified to a

Table 4.12 **Effects of eutrophication on rivers**

River zone	Effects
Headwaters with currents in the shade	None
Headwaters with currents exposed to the sun[1]	Macrophytes and periphyton growth is promoted, including filamentous algae.
Medium-sized rivers[2]	Growth of periphyton and/or macrophytes is promoted.
Large, wide rivers[3]	Growth of plankton and macrophytes.
Stagnating pools in medium size rivers	Ample growth of plankton and floating macrophytes.

Notes:
1. Average width < 1 m.
2. Average width > 1 m < 20 m, average depth < 2 m.
3. Average width > 20 m, average depth > 2 m.

great extent by human activities involving canalizing rivers or building structures to impound water. Erosion produces habitat losses followed by an impaired performance of biological communities. In addition, water movement creates turbulence and material suspension preventing sunlight penetration and photosynthetic functions. In such conditions, primary producers, the base of nutritional chains, can disappear.

Water temperature influences physiological processes. In flowing waters, a slight elevation of temperature may be beneficial for accelerating processes and organism development over the short retention times typical of such waters. Nevertheless, if the riverine temperature is increased above an acceptable level, for example by discharges of cooling waters or warm runoff, reactions stop and the organisms die.

PELAGIC COMMUNITIES

Pelagic communities are those that swim or float in rivers. Among them, only phytoplankton is able to produce large populations during the short hydraulic residence times typical of rivers, due to their high growth rates. Sunlight intensity, high temperatures, and low turbidity favour phytoplankton growth. This phenomenon also occurs where water is impounded by means of dams or other barriers that increase hydraulic residence times.

BENTHIC SPECIES

Benthic species are invertebrate organisms that live in, on or near the bottom sediments deposited on river-beds. Benthic communities are sensitive to water and sediment quality changes, and consequently are frequently used as water quality indicators.

Macrophytes can be also classified as benthic species, because they attach their roots to bottom sediments. Some macrophytes remain completely submerged while others are emergent. Since the macrophytes require light and nutrients, the changes in their composition and abundance accurately reflect the level of eutrophication or physical characteristics of the river-bed. Macrophytes constitute an important refuge for small invertebrates, fish, fry and eggs.

4.8.3.2 Lakes and reservoirs

Pollutants will disperse themselves among the various lake compartments, including water, sediment or organisms, depending on their chemical characteristics. Once a lake is contaminated, the pollutants follow diverse chemical, biological and physical pathways. The soluble compounds (hydrophilic) are transported with water and mixed in lake water during the hydraulic residence time. Phenomena like density stratification or turnover may help disperse pollutants. Non-soluble pollutants, mostly trace organic compounds and trace metals, are adsorbed to sediments. The distribution and removal of these chemicals is closely related to the sediment behaviour.

Bioaccumulation and biomagnification are important factors that influence pollution effects. There is evidence suggesting that these mechanisms produce high concentrations of polycyclic aromatic hydrocarbons (PAHs) and carcinogens in fish inhabiting waters in or near urban areas (Black, 1983). Similar effects were reported for metals.

A summary of the effects of eutrophication on biota is presented in Table 4.13. The main modification of biota caused by eutrophication is the proliferation of several kinds of aquatic weeds. Most common are water hyacinths (*Eichhornia crassipes)*, hydrilla (*Hydrilla verticillata)*, cattail (*Typha* sp.) and duckweed (*Lemna* sp.). Water hyacinths grow in a great variety of habitats – from continental pools, marshes, drains, channels, lakes and dam reservoirs to slowly flowing rivers, and adapt to a wide variety of environmental conditions. They can survive for long periods even in oligotrophic waters, but optimal growth takes place in eutrophic conditions. *Hydrilla* is probably native to the warmer regions of Asia and has spread to warm regions of the world. *Typha* sp. is a herbaceous plant that appears throughout North America, Europe and Asia, mainly in the temperate, subtropical and tropical zones. It lives along the borders of reservoirs, channels, pools and marshes, and grows densely in humid habitats or in fresh or brackish waters up to 1 m deep. *Lemna* sp. is an aquatic plant that floats on the surface of lakes, pools and bogs. Its fast vegetative propagation in

Table 4.13 **Biota status in lakes with different trophic levels**

Trophic status

Oligotrophic. Primary productivity and associated biomass is low due to low nutrient (N and P) concentrations. Fish fauna is of high quality and value. Oxygen in the water column is near the saturation point.

Mesotrophic. Fish are of intermediate quality due to oxygen deficiencies in the hypolimnion. Stratification occurs in summer.

Eutrophic. There is a significant production of biomass as a result of high nutrient concentrations. Low water transparency affects plants development. Communities of fish are of low value. The water quality eliminates or impairs many uses of water. During summer stratification, oxygen concentration in hypolimnion can be very low (<1 mg/L).

Hypereutrophic. Excessive concentrations of biomass and nutrients. Fish communities are of very low quality. Possible uses of water are very limited. Anoxic conditions exist, or there is an evident lack of oxygen in the hypolimnion during summer stratification.

Distrophic. Organic matter content is high (mainly humic and fulvic acids) and fish practically do not exist.

the aquatic environment causes high evapotranspiration rates and therefore important water losses that can be excessive in shallow water bodies.

The control of aquatic weeds consists of reducing their concentration to an acceptable level per unit area (it is impossible to eliminate them completely). Four techniques are used for such purposes: (a) biological control, (b) physical or mechanical control, (c) chemical control, and (d) manipulation of habitat.

Acidification of lakes contributes to leaching of bottom sediments and hydroxylation of iron oxides, manganese and aluminium as well as other toxic metals. Aluminium dissolution occurs at a $pH < 4.5$ and its presence is toxic to fish because it deposits as aluminium oxide on gills and causes asphyxia.

When water is withdrawn from the surface layer of reservoirs at a rate higher than the heating capacity, there is a net loss of epilimnion and primary producers are affected. However, if withdrawal is from the bottom, the cold water of the hypolimnion is withdrawn and lake water quality is improved, if no compounds are released from sediments. Water withdrawn from the bottom stratas always has a poorer chemical quality (less oxygen content and more suspended solids) than surface water, so bottom withdrawals are generally avoided.

Chapter 5

Summary

SUMMARY

Current projections of the world population growth indicate that practically the entire future increase will take place in urban areas, and more specifically, in smaller secondary cities and towns of the developing world. Maksimovic and Tejada-Guibert (2001) described the resulting demands on urban water management as one of the greatest challenges in the management of large urban entities, and such a challenge will require concerted action by the whole international community in dealing with provision of safe drinking water and improved sanitation. The material discussed in this book contributes in a distinct way to the search for, and implementation of, solutions to the urban water management problems by elucidating a basic hydrologic concept, the hydrological cycle, and its transformation in urban areas to the urban water cycle. The UWC can serve as a tool for planning ways to meet the water needs of the urban population, respond to changes in the hydrology of urban areas caused by urbanization, analyse the role of urban infrastructures in meeting urban water management needs, and mitigate the associated effects of urbanization on the environment.

The UWC provides a conceptual and unifying basis for studying the water balance of urban areas and developing water budgets that address both water quantity and quality. In its fullest, besides water balance, the UWC also addresses sediment, chemicals, micro-organisms and energy flows associated with water. The principal components of the UWC include: water inputs (mainly precipitation and municipal water supply), urban drainage and flood protection systems (including sewers, drains, streets, streams, rivers, canals, reservoirs and lakes), wastewater management systems (including the collection and transport networks, and wastewater treatment and reuse systems), and aquatic habitats. These principal components are interconnected and interdependent, with changes in any one of them usually propagating throughout the entire UWC. To meet the sometimes conflicting demands on urban water and to improve the sustainability of urban areas, modern approaches to water or natural resources management have been proposed under such terms as total management of the UWC, integrated urban water management, or simply sustainable development. In spite of this variation in terminology, all these approaches strive to achieve the same goal – meeting the needs of the urban population with respect to water in a sustainable way, while supporting economic growth and protecting the living environment. A brief summary of management concepts supporting the attainment of this goal follows.

The hydrology of urban areas is dramatically different from that of natural catchments. The changes arise from large modifications to the landscape in urban areas (increased imperviousness) and provision of denser and more hydraulically efficient drainage networks. While many of these changes are unavoidable in urban areas, their overall effect should be mitigated by emphasizing the need to maintain local water balances throughout the urban area. Specific measures connected with urban infrastructures are discussed in the following paragraph.

Three types of urban infrastructure are commonly recognized: those serving water supply, urban drainage and flood protection, and wastewater management. With respect to water supply, the primary goal in the developing world is to provide water to the large population without access to supplies of improved water (estimated at 1.1 billion people, in 2000). In developed countries, there is a shift from water supply management (based on providing increasing quantities of water) to demand management by adopting the 'soft path for water' approach. The latter approach applies to both developed and developing countries and is based on increasing the efficiency of the end water use, avoiding system water losses, and matching the supplied water quantity and quality to the requirements of the end use. In current water supply, supplementary sources of water, such as rainwater harvesting and water reclamation and reuse, are gaining in importance. The field of water supply has greatly benefited from new treatment technologies, including membranes and disinfection by UV irradiation. The safety of drinking water is assured by adopting the multiple barrier approach, managing water quality along the whole path from the source to the treatment plant, and through the distribution system to the consumer.

Urban drainage and flood protection practice has evolved dramatically during the last thirty years. With respect to drainage, the old policy of fast removal of surface runoff from urban areas has been largely abandoned and replaced by such concepts as low-impact development, sustainable urban drainage systems, or water sensitive urban design. All these concepts emphasize the need for distributed drainage systems maintaining on-site water balance and thereby reducing generation of runoff and its conveyance to downstream areas. This is achieved by minimizing imperviousness and employing green roofs, grassed drainage swales, infiltration facilities of various designs (wells, trenches and basins), storage and treatment in stormwater management ponds and constructed wetlands, and similar measures. For all these systems, design guidance and verification of performance is available. Essential issues include the overall system design, safe operation and proactive maintenance. In flood protection, the emphasis is shifting from the traditional structural or 'do nothing' alternatives to more sustainable approaches combining a full range of such measures as keeping floodplains for flood conveyance, structural measures (dams and dykes), and non-structural flood management measures, including flood mapping, zoning, insurance, and real-time flood forecasting and warning. At the same time it is recognized that in the developing world, applications of many of these measures may be limited by political and economic conditions, and institutional arrangements.

Wastewater management and sanitation is another critical issue of urban water management with respect to the monumental task of providing improved sanitation to 2.4 billion people (2000 data) who lack access to improved sanitation worldwide, and in view of the fact that most of the world's wastewaters are being discharged into the environment without any treatment. With respect to collection and treatment of wastewater, there has been considerable progress in promoting both centralized waste-water management systems (common in developed countries) and distributed systems wastewater management, which seems well suited for ecological sanitation in both developed and developing countries. Various eco-sanitation systems exist and include such steps as separation of urine from household wastewater, the processing of urine for nitrogen and phosphorus recovery and reuse in agriculture, black matter compost-ing and reuse, and simple treatment of grey water. Essential benefits of ecological san-itation include prevention of diseases, protection of the environment, return of nutrients to the soils, and affordability. With respect to treatment technologies, prom-ising approaches include natural purification systems, membrane bioreactors and anaerobic treatment systems. Such processes then facilitate wastewater reclamation and reuse, where the current concerns are of a non-technical nature (financing, failure management and social acceptability).

Chapter 4 deals with impacts of urbanization on the environment and provides a matrix listing the main physical, chemical, micro-biological and combined impacts of urbanization on six environmental compartments: the atmosphere, surface waters, wetlands, soils, groundwater and biota. In the case of the atmosphere, the most impor-tant urban impacts include the heat island phenomenon (increased temperatures), increased precipitation in downwind areas (more condensation nuclei), dry deposi-tion, polluted precipitation, and smog. In the case of surface waters, the main impacts include increased surface runoff and flooding, reduced groundwater recharge, higher water temperature, pollution of streams, rivers and lakes (including sources of drink-ing waters), and loss of biodiversity and beneficial uses. Wetlands are also affected by urbanization through changes in water balance, pollution and bacterial ecology, and loss of biodiversity. Soils are affected by increased erosion rates, soil pollution, changes in bacterial ecology due to wastewater sludge on-land applications, and landfills. The main urbanization impacts on groundwater include changes in the groundwater tables (lowering or rising), and contamination. Finally, biota is affected by loss of habitat, toxic effects, biotic impacts (diseases) and the resulting loss of biodiversity. The approaches to managing most of these risks are discussed in the chapter on urban infra-structures and elsewhere, for example in other books of the UNESCO Urban Water Management series.

In spite of many challenges, well-managed cities offer many advantages and con-tinue to attract population from rural areas. While it is essential that the existing chal-lenges of urban water management are fully recognized, it is equally important to recognize the progress taking place in developing new approaches to urban water management. In this connection, Maksimovic and Tejada-Guibert (2001) point out four broad areas of ongoing transformations in water resources management: (a) con-ceptual breakthroughs, including new paradigms in terms of ecosystems and sustain-ability, (b) methodological technological advances (multi-purpose/multi-objective approaches, computerized techniques and decision support systems, and new technolo-gies), (c) organizational mobilization with respect to new institutional arrangements and participatory processes, and (d) contextual changes, including new areas of con-cern, shifting priorities and socio-political intervention. For achieving significant progress in urban water management in the developing world, socio-economic consid-erations may be more important than technological advances. Such considerations include establishing efficient and just institutional and legal arrangements for water management appropriate to the country's degree of social and cultural conditions and economic development, finding appropriate financing of affordable water services, encouraging the effective involvement of all stakeholders, including the general public, and implementing educational programmes.

The continuing progress in our understanding of urban water management challenges, development of new approaches and technologies, public education and broad stakeholder participation, and development of new policies and institutions serving to implement such policies should contribute to meeting the current challenges in urban water management.

References

Anda, M., Ho, G.E., Mathew, K. and Monk, E. 1996. Greywater reuse options: areas for further research in Australia. J Niemczynowicz (ed.) *Environmental Research Forum: Integrated Water Management in Urban Areas*, Vol. 3. Lund, Sweden, Transtec, pp. 347–56.

Anderson, B.C., Caldwell, R.J., Crowder, A.A., Marsalek, J. and Watt, W.E. 1997. Design and operation of an aerobic biological filter for the treatment of urban stormwater runoff. *Water Qual. Res. J. Canada*, Vol. 32, No. 1, pp. 119–37.

Asano, T., Maeda, M. and Takaki, M. 1996. Wastewater reclamation and reuse in Japan: overview and implementation examples. *Wat. Sci. Tech.*, Vol. 34, No. 11, pp. 219–26.

Ashbolt, N.J. and Bruno, M. 2003. Application and refinement of the WHO risk framework for recreational waters in Sydney, Australia. *J. Water Health*, Vol. 1, pp. 125–31.

Ashley, R.M., Balmforth, D.J., Saul, A.J. and Blanksby, J.D. 2004. Flooding in the future: predicting climate change, risks and responses in urban areas. *Proceedings of the Sixth International Conference on Urban Drainage Modelling*, Dresden, 15–17 September, pp. 105–113.

Azzout, Y., Barraud, S., Cres, F.N. and Alfakih, E. 1994. *Techniques alternatives en assainissement pluvial* (in French, Alternative techniques for stormwater drainage). Paris, Lavosier Tec and Doc.

Baptista, M.B., de Oliveira Nascimento, N. and Barraud, S. 2005. *Tecnicas compensatorias em drenagem urbana* (in Portuguese, Compensation technqiues in urban drainage). Porto Alegre, Brazil, ABRH Press.

Barrios-Pérez, J. 2003. *Sludges Acid Stabilization*. PhD thesis in environmental engineering, DEPFI-UNAM, Mexico (in Spanish).

Batterman, S., Ahang, L., Wang, S. and Franzblau, A. 2002. Partition coefficients for the trihalomethanes among blood, urine, water, milk and air. *Sci. Total Envir*, Vol. 284, pp. 237–47.

Baumann, D.D., Boland, J.J. and Hanemann, W.M. 1998. *Urban Water Demand Management and Planning*. New York, McGraw-Hill.

Bausum, H., Schaub, A., Bats, R., McKim, H., Shumacher, P. and Brockett, B. 1983. Microbiological aerosols from field source wastewater irrigation system. *J. Wat. Pollut. Control Fed.* Vol. 55, No. 1, pp. 65–80.

Bedient, P.H., Rifai, H.S. and Newell, C.J. 1994. *Ground Water Contamination: Transport and Remediation*. Englewood Cliffs, N.J., Prentice Hall.

Berg, G. 1973. Reassessment of the virus problem in sewage and in surface and renovated waters. *Progress in Water Technology*, Vol. 3, pp. 87–94.

Berkoff, J. 1994. *A Strategy for Managing Water in the Middle East and North Africa*. Washington, D.C., World Bank.

Bhatia, R. and Falkenmark, M. 1993. *Water Resource Policies and the Urban Poor: innovative approaches and policy imperatives*. Washington, D.C., World Bank.

Bishop, C.A., Struger, J., Barton, D.R., Shirose, L.J., Dunn, L, Lang, A.L. and Shepherd, D. 2000a. Contamination and wildlife communities in stormwater detention ponds in Guelph and the Greater Toronto Area, Ontario, 1997 and 1998. Part I: Wildlife communities. *Wat. Qual. Res. J. Canada*, Vol. 35, No. 3, pp. 399–435.

Bishop, C.A., Struger, J., Shirose, L.J., Dunn, L. and Campbell, G.D. 2000b. Contamination and wildlife communities in stormwater detention ponds in Guelph and the Greater Toronto Area, Ontario, 1997 and 1998. Part II – Contamination and biological effects of contamination. *Wat. Qual. Res. J. Canada*, Vol. 35, No. 3, pp. 437–474.

Bixby, R. and O'Brien, J. 1979. Influence of fulvic acid on bacteriophage adsorption and complexation in soil. *Applied Envir. Microbiol*, Vol. 38, pp. 840–5.

Bixio, D., De Heyder, B., Cikurel, H., Muston, M., Miska, V., Joksimovic, D., Schaefer, A.I., Ravazzini, A., Aharoni, A., Savic D. and Thoeye, C. 2005. Municipal wastewater reclamation: where do we stand? An overview of treatment technology and management practice. *Wat. Sci. Tech.*, Vol. 5, No. 1, pp. 77–85.

Black, J.J. 1983. Field and laboratory studies of environmental carcinogenesis in Niagara River fish. *J. Great Lakes Res.*, Vol. 9, pp. 326–34.

Booth, D.B. 1990. Stream-channel incision following drainage-basin urbanization. *Water Resour. Bull.*, Vol. 26, pp. 407–17.

Borchardt, D. and Statzner, B. 1990. Ecological impact of urban stormwater runoff studied in experimental flumes: population loss by drift and availability of refugial space. *Aquatic Sciences*, Vol. 52, No. 4, pp. 299–314.

Bouwer, H. 1989. Groundwater recharge with sewage effluent. *Wat. Sci. Tech.*, Vol. 23, pp. 2099–108.

Bouwer, H. and Rice, R.C. 1984. Renovation of wastewater at the 23rd Avenue rapid infiltration project. *J. Wat. Pollut. Control Fed.*, Vol. 56, No. 1, pp. 76–83.

Bouwer, H., Rice, R., Lance, J and Gilbert, R. 1980. Rapid infiltration research at flushing meadows project Arizona. *J. Wat. Pollut. Control Fed.*, Vol. 52, pp. 2457–70.

Bratton, R. and Nesse, R. 1993. Ascariasis: an infection to watch for immigrants. *Postgrad. Med.*, Vol. 93, pp. 171–8.

Brundtland, G. (ed.) 1987. *Our Common Future*. The World Commission on Environment and Development. Oxford, Oxford University Press.

Butler, D. and Memon, F. (eds) 2005. *Water Demand Management*. London, IWA Press.

Capella, A. 2002. *Situación del Agua en el Valle de México*. Washington, D.C., World Bank internal report.

Chambers, P.A., Allard, M., Walker, S.L., Marsalek, J., Lawrence, J., Servos, M., Busnarda, J., Munger, K.S., Adare, K., Jefferson, C., Kent, R.A. and Wong, M.P. 1997. Impacts of municipal wastewater effluents on Canadian waters: a review. *Wat. Qual. Res. J. Canada*, Vol. 32, No. 4, pp. 659–713.

Chapman, D. (ed.) 1992. *Water Quality Assessment*. WHO/UNESCO/UNEP. London, Chapman and Hall.

Chávez, A., Jiménez, B., and Gilberg, L. 2002. Particle and micro-organism removal in treated sewage using several types of coagulants. H.H. Hoffmann and H. Odegaard (eds) *Chemical Water and Wastewater Treatment: 10th Gothenburg Symposium*. London, IWA Press., pp. 213–21.

Che, W., Liu, C., Liu, Y., Li, J. and Wang, W. 2004. First flush control of urban runoff pollution. CD-ROM *Proceedings of the 8th International Conference on Diffuse/Nonpoint Pollution*, Kyoto, 24–29 October 2004.

Cheremisinoff, P. 1992. *A Guide to Underground Storage Tanks: evaluation, site assessment and remediation*. Englewood Cliffs, N.J., Prentice Hall.

Chocat, B. (ed.) 1997. *Encyclopédie de l'hydrologie urbaine et de l'assainissement* (in French, Encyclopedia of urban hydrology and sanitation). Paris, Lavosier Tec and Doc.

Chocat, B. 2006. Personal communication.

Chow, V.T. 1964. *Handbook of Applied Hydrology*. New York, McGraw-Hill.

Cifuentes, E., Suárez, L., Solano, M. and Santos, R. 2002. Diarrhea diseases in children from a water reclamation site, Mexico City. *Environmental Health Perspectives*, Vol. 110, No. 10, pp. A619–24.

Cockram, M. and Feldman, S. 1996. The beautiful city gardens in third-world cities. *African Urban*, Vol. 11, Nos. 2–3, pp. 202–8.

Cohen, B. 2006. Urbanization in developing countries: current trends, future projections, and key challenges for sustainability. *Technology in Society*, Vol. 28, pp. 63–80.

Colas, H., Pleau, M., Lamarre, J., Pelletier, G. and Lavalle, P. 2004. Practical perspective on real-time control. *Wat. Qual. Res. J. Canada*, Vol. 39, No. 4, pp. 466–78.

Commission of the European Communities (CEC) 2002. *Water Management in Developing Countries: policy and priorities for EU development cooperation*. Brussels, CEC, COM.

Corbitt, R. (ed.) 1990. *Standard Handbook of Environmental Engineering*. New York, McGraw-Hill.

Council of European Communities 1991. *The Urban Waste Water Directive 91/271/EC*. EU, Brussels, Belgium (also available on the Internet).

Council of European Communities 1998. *The Drinking Water Directive 98/83/EC*. EU, Brussels, Belgium (also available on the Internet).

Cowden, J.R., Mihelcic, J.R. and Watkins, D.W. 2006. Domestic rainwater harvesting assessment to improve water supply and health in Africa's urban slums. *Proceedings of World Environmental and Water Resources Congress*, May, 2006, Omaha, Nebr.

Craun, G. 1984. Health aspects of groundwater pollution. G Bitton and C.P. Gerba (eds) *Groundwater Pollution Microbiology*. New York, John Wiley, pp. 135–79.

Crook, J., Herndon, R.L., Wehner, M.P. and Rigby, M.G. 1995. Studies to determine the effects of injecting 100 percent reclaimed water from Water Factory 21. *Conference Proceedings of the Water Environment Federation Annual Technical Exhibition and Conference (WEFTEC) '95, 68th Annual conference*, Vol. 6, Miami Beach, Fla.

Crowther, R.A. and Hynes, H.B.N. 1977. Effect of rural highway runoff on stream benthic macrovertebrates. *Environ. Poll. Series A*, Vol. 32, pp. 157–70.

David, L.M. and Matos, J.S. 2005. Wet-weather urban discharges: implications from adopting the revised European Directive concerning the quality of bathing water. *Wat. Sci. Tech.*, Vol. 52, No. 3, pp. 9–17.

Department of Irrigation and Drainage (DID) 2001. *Urban Stormwater Management Manual*. Putrajaya, Malaysia, Department of Irrigation and Drainage.

Desbordes, M. and Servat, E. 1988. Towards a specific approach of urban hydrology in Africa. *Proceedings of the International Symposium on Hydrologic Processes and Water Management in Urban Areas, Urban Water 88*, Duisburg, 24–29 April, pp. 231–7.

Dick-Peddie, W.A. 1991. Semiarid and arid lands: a worldwide scope. J. Skujins (ed.) *Semiarid Lands and Deserts: soil resource and reclamation*. New York, Marcel Dekker, pp. 3–32.

Drewry, W. and Eliassen, R. 1968. Virus movement in groundwater. *J. Wat. Pollut. Control Fed.*, Vol. 40, pp. 257–71.

Duncan, H.P. 1999. *Urban Stormwater Quality: a statistical overview*. Melbourne, Australia, Cooperative Research Centre for Catchment Hydrology, Report 99/3.

Dunne, T. 1986. Urban hydrology in the tropics: problems solutions, data collection and analysis. *Urban Climatology and Its Application with Special Regards to Tropical Areas, Proceedings of the Mexico Tech Conference*, November 1984, World Climate Programme. Geneva, WMO.

Dutka, B.J. and Marsalek, J. 1993. Urban impacts on river shoreline microbiological pollution. *J. Great Lakes Res.*, Vol. 19, No. 4, pp. 665–74.

Dutka, B.J., Marsalek, J., Jurkovic, A., McInnis, R. and Kwan, K.K. 1994a. Ecotoxicological study of stormwater ponds under winter conditions. *Zeitschrift für angewandte Zoologie*, Vol. 80, No. 1, pp. 25–42.

Dutka, B.J., Marsalek, J., Jurkovic, A., McInnis, R. and Kwan, K.K. 1994b. A seasonal ecotox-icological study of stormwater ponds. *Zeitschrift für angewandte Zoologie*, Vol. 80, No. 3, pp. 361–81.

EcoSanRes 2005. Fact sheet 5: http://www.ecosanres.org/PDF%20files/Fact_sheets/ESR5 lowres.pdf

Edge, T.A. and Schaefer, K. (eds) 2006. *Microbial Source Tracking in Aquatic Ecosystems: the state of science and an assessment of needs.* NWRI Scientific Assessment Report Series No. 7 and Linking Water Science to Policy Workshop Series. Burlington, Ontario, National Water Research Institute.

Ensink, J., Simmons, J. and van der Hoek, W. 2004a. Wastewater use in Pakistan: the cases of Haroonabad and Faisalabad. C. Scott, N. Faruqui and L. Raschid-Sally (eds) *Wastewater Use in Irrigated Agriculture: confronting the livelihood and environmental realities.* Wallingford, UK, CAB International.

Ensink, J., Mahmood, T., van der Hoek, W., Raschid-Sally, L. and Amerasinghe, F. 2004b. A nationwide assessment of wastewater use in Pakistan: an obscure activity or a vitally impor-tant one? *Water Policy*, Vol. 6, pp. 197–206.

Esrey, S.A., Andersson, I., Hillers, A. and Sawyer, R. 2001. *Closing the Loop: ecological sanitation for food security.* UNDP Publications on Water Resources No. 18. New York, UNDP.

European Environment Agency (EEA) 1999. *Environment in the European Union at the Turn of the Century.* Environmental assessment report No. 2. Copenhagen, EEA.

Falkland, A. (ed.) 1991. *Hydrology and Water Resources of Small Islands: a practice guide.* Paris, UNESCO.

Fernandez-Breaumontz, A. 1988. Contingencies and quality of the air in the City of Mexico (in Spanish). *Bulletin of Development and Investigation on the Quality of the Air in Great Cities*, Year I, No. 1, July–December, pp. 2–3.

Fetter, C. 1988. *Applied Hydrogeology*, 2nd edn. New York, Merrill Publishing Company.

Figueroa Vega, G.E. 1984. Case history No. 9.8 Mexico, D.F., Mexico. J.F. Poland (ed.) *Guidebook to Studies of Land Subsidence due to Ground-water Withdrawal.* Studies and Reports in Hydrology 40, prepared for the International Hydrological Programme, Working Group 8.4. Paris, UNESCO, pp. 217–32.

Foresti, E. 2001. Perspectives on anaerobic treatment in developing countries. *Wat. Sci. Tech.*, Vol. 44, No. 8, pp. 141–8.

Foster, S. 2001. Groundwater recharge with urban wastewater reconciling resource recovery and pollution concerns in developing nations. Paper presented in the *Experts Meeting on Health Risks in Aquifer Recharge by Reclaimed Water*, 9–10 November 2001, Budapest, Hungary sponsored by WHO.

Foster, S., Hirata, R., Gomes, D., D'Elia, M. and Paris, M. 2002. *Groundwater Quality Protection: a guide for water service companies, municipal authorities and environment agen-cies.* World Bank Group and Global Water Partnership (eds). Washington, D.C., World Bank.

Frederiksen, H.D., Berkoff, J. and Barber, W. 1993. *Water Resources Management in Asia, Volume I.* Washington, D.C., World Bank, Technical Paper 212.

Fujita, Y., Ding, W.H. and Reinhard, M. 1996. Identification of wastewater dissolved organic car-bon characteristics in recycled wastewater and recharge groundwater. *Wat. Env. Res.*, Vol. 68, No. 5, pp. 867–76.

Galli, F.J. 1991. *Thermal Impacts Associated with Urbanization and BMPs in Maryland.* Washington, D.C., Metropolitan Washington Council of Governments.

Geake, A.K. and Foster, S.D. 1986. *Groundwater Recharge Controls and Pollution Pathways in the Alluvial Aquifer of Metropolitan Lima, Peru: results of detailed research at La Molina Alta and at San Juan de Miraflores wastewater reuse complex.* Wallingford, UK, British Geological Survey.

Geiger, W.F., Marsalek, J., Rawls, W.J. and Zuidema, F.C. (eds) 1987. *Manual on Drainage in Urban Areas. Volume I: Planning and design of drainage systems.* Studies and reports in hydrology No. 43. Paris, UNESCO.

Gerba, C. and Bitton, G. 1984. Microbial pollutants: their survival and transport pattern to groundwater. G. Bitton and C. Gerba (eds) *Groundwater Pollution Microbiology.* New York, John Wiley, Chapter 4.

Gersburg, R., Gerhart, R. and Ives, M. 1989. Pathogen removal in constructed wetlands. Hammer, D. (ed.) *Constructed Wetlands for Wastewater Treatment: municipal, industrial and agricultural.* Chelsea, Mich., Lewis Publishers, pp. 431–45.

Ghosh, D. 1999. *Wastewater Utilization in East Calcutta.* Gouda, the Netherlands, Urban Waste Expertise Program (UWEP) Occasional paper.

Gibbons, J. and Laha, S. 1999. Water purification systems: comparative analysis based on the occurrence of disinfection by-products. *Envir. Pollut.*, Vol. 106, pp. 425–8.

Girovich, M. 1996. *Biosolids Treatment and Management Process for Beneficial Use.* New York, Marcel Dekker.

Gleick, P.H. 1998. *The World's Water 1998–1999: the biennial report on freshwater resources.* Washington, D.C., Island Press.

Goldman, J.M. and Murr, A.S. 2002. Alterations in ovarian follicular progesterone secretion by elevated exposures to the drinking water or the potential site(s) of impact along the steroidogenic pathway. *Toxicology*, Vol. 171, pp. 83–93.

Goldstein, S.T., Juranek, D.D., Ravenholt, O., Hightower, A.W., Martin, D.G., Mesnik, J.L., Griffiths, S.D., Bryant, A.J., Reich, R.R. and Herwaldt, B.L. 1996. Cryptosporidiosis: an outbreak associated with drinking water despite state-of-the-art water treatment. *Ann. Intern. Med.*, Vol. 124, No. 5, pp. 459–68.

Gray, N. 1994. *Drinking Water Quality*, Chichester, UK, John Wiley, pp. 150–70.

Grimmond, C.S.B. and Oke, T.R. 1986. Urban water balances 2: results from a suburb of Vancouver, British Columbia. *Wat. Resour. Res.*, Vol. 22, No. 10, pp. 1404–12.

Grum, M., Jørgensen, A.T., Johansen, R.M. and Linde, J.J. 2006. The effects of climate change in urban drainage: an evaluation based on regional climate model simulations. *Wat. Sci. Tech.*, Vol. 54, Nos. 6–7, pp. 9–16.

Gumbo, B. 2000. Mass balancing as a tool for assessing integrated urban water management. *Proceedings of the 1st WARFSA/WaterNet Symposium*, Maputo, Mozambique, Nov. 1–2, 2000, pp. 1–10.

Gunther, F. 2000. Wastewater treatment by greywater separation: outline of a biologically based greywater purification plant in Sweden. *Ecological Engineering*, Vol. 15, pp. 139–46.

Hall, K.J. and Anderson, B.C. 1988. The toxicity and chemical composition of urban stormwater runoff. *Can. J. Civil Eng.*, Vol. 15, pp. 98–105.

Hammer, M.J. and Hammer, M.J., Jr. 2003. *Water and wastewater technology*, 5th edn. Englewood Cliffs, N.J., Prentice Hall.

Haramoto, E., Katayama, H., Oguma, K., Koibuchi, Y., Furumai, H. and Ohgaki, S. 2006. Effects of rainfall on the occurrence of human adenoviruses, total coliforms, and Escherichia coli in seawater. *Wat. Sci. Tech.*, Vol. 54, No. 3, pp. 225–30.

Harremoes, P. 1988. Stochastic models for estimation of extreme pollution from urban runoff. *Wat. Res.*, Vol. 22, pp. 1017–26.

Hashimoto, T., Stedinger, J.R. and Loucks, D.P. 1982. Reliability, resiliency, and vulnerability criteria for water resource system performance evaluation. *Water Resour. Res.*, Vol. 18, No. 1, pp. 14–20.

Hauger, M. B., Mouchel, J.M. and Mikkelsen, P.S. 2006. Indicators of hazard, vulnerability and risk in urban drainage. *Wat. Sci. Tech.*, Vol. 54, No. 6–7, pp. 441–50.

Henry, J.G. and Heinke, G.W. 1989. *Environmental science and engineering.* Prentice-Hall, Englewood Cliffs, N.J.

Herrmann, T. and Klaus, U. 1996. Fluxes of nutrients in urban drainage systems: assessment of sources, pathways and treatment techniques. F. Sieker and H.-R. Verworn (eds) *Proceedings of the 7th International Conference on Urban Storm Drainage*, Hannover, Germany, 9–13 September 1996, pp. 761–6.

Hiley, P.D. 1995. The reality of sewage treatment using wetlands, *Wat. Sci. Tech.*, Vol. 32, No. 3, pp. 329–38.

Ho, G. 2004. Small water and wastewater systems: pathways to sustainable development? *Wat. Sci. Tech.*, Vol. 48, Nos. 11–12, pp. 7–14.

Ho, G. 2005. Technology for sustainability: the role of onsite, small and community scale technology. *Wat. Sci. Tech.*, Vol. 51, No. 10, pp. 15–20.

Hogg, D. 2002. *Cost for Municipal Waste Management in the UE*. A report prepared by Eunomia Research and Consulting to the Directorate General Environment., Brussels, Belgium, European Commission.

Hogland, W. and Niemczynowicz, J. 1980. *Kvantitativ och kvalitativ vattenom-sattningsbudget for Lund centralort; kompletterande matning gar och methodic* (in Swedish; Water quantity and quality budget for the city of Lund: estimation and methodology). Report 3038, Dept. of Water Resources, University of Lund, Sweden.

Hopkins, R., Vial, P., Ferrecio, C., Ovalle, J., Prado, P., Sotomayor, V., Russell, R., Wasserman, A. and Morris, J. 1993. Seroprevalence of *Helicobacter pylori* in Chile: vegetables may serve as one route of transmission. *J. of Infectious Diseases*, Vol. 168, pp. 222–6.

Horner, R.R., Skupien, J.J., Livingston, E.H. and Shaver, H.E. 1994. *Fundamentals of Urban Runoff Management: technical and institutional issues*. Washington, D.C., Terrene Institute.

Houghton, J.T., Filho, L.G.M., Callander, B.A., Harris, N., Kattenberg, A. and Maskell, K. 1996. *Climate Change 1995: the science of climate change*. Intergovernmental Panel on Climate Change. Cambridge, UK, Cambridge University Press.

Hvitved-Jacobsen, T. 1982. Impact of combined sewer overflows on dissolved oxygen concentration of a river. *Wat. Res.*, Vol. 16, pp. 1099–1105.

Imbe, M., Ohta, T. and Takano, N. 1997. Quantitative assessment of improvements in hydrological water cycle in urbanized river basin. *Wat. Sci. Tech.*, Vol. 36, No. 8–9, pp. 219–22.

Indian Road Congress (IRC) 1999. *Guidelines on urban drainage*. New Delhi, IRC.

Jawetz, E., Melnick, J. and Adelberg, E. 1996. *Medical Microbiology, Modern Manual* (Microbiologica medica, manual moderno – in Spanish), 15th edn. Mexico City, Mexico, Manual Moderno.

Jayakumar, K.V. and Dandigi, M.N. 2003. A cost effective environmentally friendly treatment of municipal wastewater using constructed wetlands for developing countries. *World Water and Environmental Resources Congress*, Philadelphia. New York, ASCE, pp. 3521–31.

Jiménez, B. and Asano, T. 2004. Acknowledge all approaches: the global outlook on reuse. *Water*, No. 21, December, pp. 32–7.

Jiménez, B. and Chávez, A. 1998. Removal of helminth eggs in an advanced primary treatment with sludge blanket. *Environ. Technol.*, Vol. 19, pp. 1061–71.

Jiménez, B. and Chávez, A. 2004. Quality assessment of an aquifer recharged with wastewater for its potential use as drinking source: El Mezquital Valley case. *Wat. Sci. Tech.*, Vol. 50, No. 2, pp. 269–76.

Jiménez, B., Chávez, A., Maya, C. and Jardines, L. 2001. Removal of micro-organisms in different stages of wastewater treatment for Mexico City. *Wat. Sci. Tech.*, Vol. 43, No. 10, pp. 155–62.

Jiménez, B., Barrios, J., Mendez, J. and Díaz, J. 2004a. Sustainable sludge management in developing countries. *Wat. Sci. Tech*, Vol. 49, No. 10, pp. 251–8.

Jiménez, B., Mendes, J., Barrios, J., Salgado, G. and Sheibaum, C. 2004b. Characterization and evaluation of potential reuse options for wastewater sludges and combined sewer system sediments in Mexico. *Wat. Sci. Tech.*, Vol. 49, No. 10, pp. 171–8.

Jiménez, B., Austin, A., Cloete, E. and Phasha, C. 2006. Using Ecosans' sludges for crop production. *Wat. Sci. Tech.*, Vol. 54, No. 5, pp. 169–177.

Jiménez, B., Austin, A., Cloete, E., Phasha, C. and Beltran, N. 2007. Biological risks to food crops fertilized with Ecosan sludge. *Wat. Sci. Tech.*, Vol. 55, No. 7, pp. 21–29.

Johnson, A., Rice, E. and Reasoner, D. 1997. Inactivation of *Helicobacter pylori* by chlorination. *Applied and Environmental Microbiology*, Vol. 63, pp. 4969–70.

Jönsson, H., Baky, A., Jeppsson, U., Hellström, D. and Kärrman, E. 2005. Composition of urine, faeces, greywater and bio-waste for utilisation in the URWARE model. *Urban Water Report 2005: 6*. Goteborg, Chalmers University of Technology.

Judd, J.H. 1970. Lake stratification caused by runoff from street deicing. *Wat. Res.*, Vol. 4, pp. 521–32.

Kamizoulis, G. 2005. The new draft WHO guidelines for water reuse in agriculture. Tech. Workshop on *The Integration of reclaimed water in water resource management*, October, Lloret de Mar, Costa Brava, Girona, Spain.

Karamouz, M., Szidarovszki, F. and Zahraie, B. 2003. *Water resources systems analysis*. Boca Raton, Fla., Lewis Publishers.

Karamouz, M., Kerachian, R. and Zahraie, B. 2004. Monthly water resources and irrigation planning: case study of conjunctive use of surface and groundwater. *ASCE J. Irrigation and Drainage Engineering*, Vol. 130, No. 5, pp. 391–402.

Karamouz, M., Kerachian, R. and Moridi, M. 2005a. Conflict resolution in water pollution control in urban areas: a case study. *Proceedings of the 4th International Conference on Decision Making in Urban and Civil Engineering*, Porto, Portugal, 28–30 October.

Karamouz, M., Moridi, A. and Nazif, S. 2005b. Development of an object oriented programming model for water transfer in Tehran Metropolitan Area. *Proceedings of the 10th International Conference on Urban Drainage*, Copenhagen, Denmark, 21–26 August.

Karpiscak, M.M., Gerba, C.P., Wat, P.M., Foster, K.E. and Falabi, J.A. 1996. Multi-species plant systems for wastewater quality improvements and habitat enhancement. *Wat. Sci. Tech.*, Vol. 33, Nos. 10–11, pp. 231–6.

Kinzelman, J., McLellan, S.L., Daniels, A.D., Cashin, S., Singh, A., Gradus, S. and Bagley, R. 2004. Non-point source pollution: determination of replication versus persistence of *Escherichia coli* in surface water and sediments with correlation of levels to readily measurable environmental parameters. *J. of Water and Health*, Vol. 2, pp. 103–14.

Korn, C., Andrew, R.C. and Escobar, M.D. 2002. Development of chlorine dioxide-related by-product models for drinking water treatment. *Wat. Res.*, Vol. 36, No. 1, pp. 330–42.

Kostecki, P.T. and Calabrese, E. 1989. *Petroleum Contaminated Soils: remediation techniques, environmental fate and risk assessment*, Vol. 1. Chelsea, Mich., Lewis Publishers.

Lance, J.C. and Gerba, C.P. 1980. Poliovirus movement during high rate land filtration of sewage water. *J. Environ. Qual.*, Vol. 9, pp. 31–4.

Lawrence, A.I., Ellis, J.B., Marsalek, J., Urbonas, B. and Phillips, B.C. 1999. Total urban water cycle based management. I.B. Joliffe and J.E. Ball (eds) *Urban Storm Drainage, Proceedings of the 8th International Conference on Urban Drainage*, Vol. 3, pp. 1142–9.

Leclerc, H., Schwartzbrod, L. and Dei-Cas, E. 2002. Microbial agents associated with waterborne diseases. *Crit. Rev. Microbiol.*, Vol. 28, No. 4, pp. 371–409.

Lefevre, N.M. and Lewis, G.D. 2003. The role of resuspension in enterococci distribution in water at an urban beach. *Wat. Sci. Tech.*, Vol. 47, No. 3, pp. 205–10.

Leopold, L. 1968. *Hydrology for Urban Land Planning: a guidebook on the hydrologic effects of urban land use*. US Geol. Survey Circular 554. Reston, Va., USGS.

Lerner, D.N. (ed.) 2004. *Urban groundwater pollution*. Lisse, the Netherlands, Balkema.

Lewis, G., Austin, J., Loutit, M. and Sharples, K. 1986. Enterovirus removal from sewage: the effectiveness of four different treatment plants. *Wat. Res.*, Vol. 20, No. 10, pp. 1291–97.

Li, Z. and Geiger, W.F. 2006. Perspectives and hindrances for on-site rainwater and greywater utilisation system in residential areas in China. CD-ROM, *Proceedings of the IWA World Water Congress and Exhibition*, Beijing, China, 10–14 September.

Lijklema, L., Roijackers, R.M. and Cupper, J.G.M. 1989. Biological assessment of effects of CSOs and stormwater discharges. J.B. Ellis (ed.) *Urban discharges and receiving water quality impacts*. (Adv. Wat. Poll. Control No.7). Oxford, Pergamon, pp. 37–46.

Lijklema, L., Tyson, J.M. and Lesouf, A. 1993. Interactions between sewers, treatment plants and receiving waters in urban areas: a summary of the INTERURBA '92 workshop conclusions. *Wat. Sci. Tech.*, Vol. 27, No. 12, pp. 1–29.

Lima, A. and Lima, N. 1993. Epidemiology, therapy and prevention of infection with *Shigella* organisms and *Clostridium difficile*. *Curr. Op. Microbiol. Infect. Dis.*, Vol. 6, pp. 63–71.

Lins, H.F. and Stakhiv, E.Z. 1998. Managing a nation's water in a changing climate. *J. of the AWRA*, Vol. 34, pp. 1255–64.

Lohani, B.N. 2005. Advancing sanitation and wastewater management agenda in Asia and Pacific Region. *Sanitation and Wastewater Management: The Way Forward Workshop*, 19–20 September 2005, Manila, Philippines.

Lue-Hing, C., Zenz, D.R. and Kuchenrither, R. (eds) 1992. *Municipal Sewage Sludge Management: processing, utilization and disposal*. Lancaster, Pa., Technomic Publishing Company.

Lundqvist, J., Narain, S. and Turton, A. 2001. Social, institutional and regulatory issues. C. Maksimovic and J.A. Tejada-Guibert (eds) *Frontiers in Urban Water Management: deadlock or hope?* London, IWA Press, pp. 344–98.

Maidment, D.R. (ed.) 1993. *Handbook of Hydrology*. New York, McGraw-Hill.

Makepeace, D.K., Smith, D.W. and Stanley, S.J. 1995. Urban stormwater quality: summary of contaminant data. *Crit. Rev. Environ. Sci. Technol.*, Vol. 25, pp. 93–139.

Maksimovic, C. and Tejada-Guibert, J.A. (eds) 2001. *Frontiers on Urban Water Management: deadlock or hope?* London, IWA Press.

Mara, D. and Cairncross, S. 1989. *Guidelines for the Safe Use of Wastewater and Excreta in Agriculture and Aquaculture*. Geneva, World Health Organization.

Marsalek, J. 1980. Runoff from developing areas. *Technical Workshop Series No. 2*, Ottawa, Canada, Inland Waters Directorate, pp. 266–76.

Marsalek, J. 1992. Overview of sediment issues in urban drainage. *Proceedings of the International Symposium on Urban Stormwater Management*, Sydney, 4–7 February, pp. 30–7.

Marsalek, J. 2003a. Overview of urban stormwater impacts on receiving waters. R. Arsov, J. Marsalek, W.E. Watt and E. Zeman (eds) *Urban water management*. NATO Science Series Vol. 25. Dordrecht, Kluwer Academic, pp. 1–10.

Marsalek, J. 2003b. Road salts in urban stormwater: an emerging issue in stormwater management in cold climates. *Wat. Sci. Tech.*, Vol. 48, No. 9, pp. 61–70.

Marsalek, J., and Rochfort, Q. 1999. Toxicity of urban wet-weather pollution sources: stormwater and CSOs. I.B. Jollife and J.E. Ball, (eds) *Proceedings of the 8th International Conference on Urban Storm Drainage*, Vol. 4, pp. 1575–82.

Marsalek, J., Barnwell, T.O., Geiger, W. F., Grottker, M., Huber, W.C., Saul, A.J., Schilling, W. and Torno, H.C. 1993. Urban drainage systems: design and operation. *Wat. Sci. Tech.*, Vol. 27, No. 12, pp. 31–70.

Marsalek, J., Dutka, B.J. and Tsanis, I.K. 1994. Urban impacts on microbiological pollution of the St.Clair River in Sarnia, Ontario. *Wat. Sci. Tech.*, Vol. 30, No. 1, pp. 177–84.

Marsalek, J., Rochfort, Q., Brownlee, B., Mayer, T. and Servos, M. 1999a. An exploratory study of urban runoff toxicity. *Wat. Sci. Tech.*, Vol. 39, No. 12, pp. 33–9.

Marsalek, J., Rochfort, Q., Mayer, T., Servos, M., Dutka, B.J. and Brownlee, B. 1999b. Toxicity testing for controlling urban wet-weather pollution: advantages and limitations. *Urban Water*, Vol. 1, No. 1, pp. 91–104.

Marsalek, J., Rochfort, Q. and Savic, D. 2001. Urban water as a part of integrated catchment management. C. Maksimovic and J.A. Tejada-Guibert (eds) *Frontiers on Urban Water Management: deadlock or hope?* London, IWA Press, Chapter 2, pp. 37–83.

Marsalek, J., Oberts, G., Exall, K. and Viklander M. 2003. Review of operation of urban drainage systems in cold weather: water quality considerations. *Wat. Sci. Tech.*, Vol. 48, No. 9, pp. 11–20.

Marsalek, P.M. 1997. *Special Characteristics of an On-stream Stormwater Pond: winter regime and accumulation of sediment and associated contaminants.* M.Sc. thesis, Dept. of Civil Engineering, Queen's University, Kingston, Ontario, Canada.

Marshall, J. 2005. Megacity, mega mess. *Nature*, September 15, pp. 312–14.

Matsui, S., Takigami, H., Matsuda, T., Taniguchi, N., Adachi, J., Kawami, H. and Shimizu, Y. 2000. Estrogen and estrogen mimics contamination in water and the role of sewage treatment. *Wat. Sci. Tech.*, Vol. 42, No. 12, pp. 173–9.

Mays, L.W. (ed.) 1996. *Water Resources Handbook*. New York, McGraw-Hill.

McPherson, M.B. 1973. Need for metropolitan water balance inventories. *J. of Hyd. Div., ASCE*, Vol. 99, No. HY10, pp. 1837–48.

McPherson, M.B. and Schneider, W.J. 1974. Problems in modeling urban watersheds. *Wat. Res.*, Vol. 10, No. 3, pp. 434–40.

Metcalf & Eddy, Inc. 2003. *Wastewater Engineering, Treatment And Reuse*, 4th edn. New York, McGraw-Hill.

Meybeck, M. and Helmer, R. 1989. The quality of rivers: from pristine state to global pollution. *Paleogeo. Paleoclimat. Paleoecol.*, Vol. 75, pp. 283–309.

Miller, D. and Scalf, M. 1974. New priorities for groundwater quality protection. *Ground Water*, Vol. 12, pp. 335–47.

Ministry of the Environment (MOE) 2003. *Stormwater Management Planning and Design Manual*. Toronto, Ontario Ministry of the Environment.

Molina, L.T. and Molina, M.J. (eds) 2002. *Air Quality in the Mexico Megacity: an integrated assessment*. Boston, Mass., Kluwer Academic.

Myer, W.B. 1991. Urban heat island and urban health: early American perspective. *Professional Geographer*, Vol. 43, No. 1, pp. 38–48.

Nachamkin, I. 1993. *Campylobacter* infections. *Curr. Op. Microbiol. Infect. Diseases*, Vol. 6, pp. 72–6.

National Research Council (NRC) 1994. *Alternatives to Groundwater Cleanup*. Washington, D.C., National Academy Press.

Nestor I. and Costin, L. 1971. Removal of the cocksakie virus from water by sand obtained from rapid sand filters of water plants. *J. of Hygiene Epidemiology Microbiology and Immunology*, Vol. 15, pp. 129–36.

New South Wales (NSW) Health 2000. *Greywater Reuse in Severed Single Domestic Premises*. North Sydney, Australia, New South Wales Department of Health.

Nhapi, I., Siebel, M.A. and Gijzen, H.J. 2006. A proposal for managing wastewater in Harare, Zimbabwe. *Water and Environment J.*, Vol. 20, No. 2, pp. 101–8.

Nicholson, R.V., Cherry, J.A. and Reardon, E.J. 1983. Migration of contaminants in groundwater at a landfill: a case study. *Journal of Hydrology*, Vol. 63, pp. 131–76.

Norris, R.D., Hinchee, R.E., Brown, R., McCarty, P.L., Semprini, L., Wilson, J.T., Kampbell, D.H., Reinhard, M., Bouwer, E.J., Borden, R.C., Vogel, T.M., Thomas, J.M. and Ward, C.H. (eds) 1994. *Handbook of Bioremediation*. Boca Raton, Fla., CRC Press and Lewis Publishers.

Notodarmojo, S., Sutrisno, R. and Simanjuntak, R.I. 2004. Characteristics of runoff from two residential areas in the city of Bandung, Indonesia. CD-ROM, *Proceedings of the 8th International Conference on Diffuse/Nonpoint Pollution*, Kyoto, 24–29 October.

Novotny, V., Muehring, D., Zitomer, D.H., Smith, D.W. and Facey, R. 1998. Cyanide and metal pollution by urban snowmelt: impact of deicing compounds. *Wat. Sci. Tech.*, Vol. 38, No. 10, pp. 223–30.

Novotny, V. and Olem, H. 1994. *Water Quality: prevention, identification and management of diffuse pollution*. New York, Van Nostrand Rheinhold.

Olivieri, V.P., Kawata, K. and Lim, S.H. 1989. Microbiological impacts of storm sewer overflows. J.B. Ellis (ed.) *Urban Discharges and Receiving Water Quality Impacts*. (Adv. Wat. Poll. Control No.7). Oxford, Pergamon, pp. 47–54.

Oron, G. 2001. *Management of Effluent Reclamation via Soil-Aquifer-Treatment Procedures*. WHO Expert consultation on Health Risks in Aquifer Recharge by Recycled Water. Budapest, 8–9 November.

Overton, E.D. and Meadows, M.E. 1976. *Stormwater Modeling*. New York, Academic Press.

Parkinson, J. 2002. Stormwater management and urban drainage in developing countries. *Intermediate Technology Publishing*, Vol. 20, No. 4, available at: http://www.sanicon.net/titles/topicintro.php3?topicId=5

Parkinson, J. and Mark, O. 2005. *Urban Stormwater Management in Developing Countries*. London, IWA Press.

Perfler, R. and Haberl, R. 1993. Actual experiences with the use of reed bed systems for wastewater treatment in single households. *Wat. Sci. Tech.*, Vol. 28, No. 10, pp. 141–8.

Pontius, F.W. (ed.) 1990. *Water Quality and Treatment: a handbook of community water supplies*, 4th edn. New York, McGraw-Hill.

Prosad, T., Kumar, B.S. and Kumar, S. 1987. Water resources development in India: its central role in the past and crucial significance for the future. *International Symposium on Water for the Future*, Rome, April, pp. 19–34.

Pyne, R.D.G. 1995. *Groundwater Recharge and Wells: a guide to aquifer storage and recovery*. Boca Raton, Fla., CRC Press/Lewis Publishers.

Qian, Y. 2000a. Appropriate technologies for municipal wastewater treatment in China. *J. Environmental Science and Health, Part A: Toxic/Hazardous Substances and Environmental Engineering*, Vol. 35, No. 10, pp. 1749–60.

Qian, Y. 2000b. Appropriate process and technology for wastewater treatment and reclamation in China. *Wat. Sci. Tech.*, Vol. 42, No. 12, pp. 107–14.

Quanrud, D.M., Arnold, R.G., Wilson, L.G., Gordon, H., Graham, D. and Amy, G. 1996. Soil Fate of organics during column studies of aquifer treatment, *J. Envir. Eng.*, Vol. 122, pp. 314–21.

Raghavendra, S. 2006. Re-examining the 'low water tariff' hypothesis: lessons from Hyderabad, India. *Urban Water*, Vol. 3, No. 4, pp. 235–47.

Raghupati, U. and Foster, V. 2002. Water tariffs and subsidies in South Asia: understanding the basics, a score card for India. *Water and Sanitation Program Paper No. 21*. Washington, D.C., World Bank.

Ramirez, E., Robles, E., Bonilla, P., Sainz, G., Lopez, M., De La Cerda, J.M. and Warren, A. 1993. Occurrence of pathogenic free-living amoebae and bacterial indicators in a constructed wetland treating domestic wastewater from a single household. *Eng. in Life Sci.*, Vol. 5, No. 3, pp. 253–8.

Richardson, C.J., and McCarthy, E.J. 1994. Effect of land development and forest management on hydrologic response in south-eastern coastal wetlands: a review. *Wetlands*, Vol. 14, pp. 56–71.

Riordan, E.J., Grigg, N.S. and Hiller, R.L. 1978. Measuring the effects of urbanization on the hydrologic regimen. P.R. Helliwell (ed.) *Urban Drainage, Proceedings of the International Conference on Urban Storm Drainage, Southampton*, April 1978, pp. 496–511.

Rishel, J.B. 2002. *Water Pumps and Pumping Systems*. New York, McGraw-Hill.

Rivera, F., Warren, A., Ramirez, E., Decamp, O., Bonilla, P.O., Gallegos, E., Calderon, A. and Sanchez, J.T. 1995. Removal of pathogens from wastewater by the root zone method (RZM). *Wat. Sci. Tech.*, Vol. 32, No. 3, pp. 211–18.

Rivera, F., Warren, A., Curds, C.R., Robles, E., Gutierrez, A., Gallegos, E. and Calderon, A. 1997. The application of the root zone method for the treatment and reuse of high-strength abattoir waste in Mexico. *Wat. Sci. Tech.*, Vol. 35, No. 5, pp. 271–8.

Rochfort, Q., Anderson, B.C., Crowder, A.A., Marsalek, J. and Watt, W.E. 1997. Field-scale studies of subsurface flow constructed wetlands for stormwater quality enhancement. *Water Qual. Res. J. Canada*, Vol. 32, No. 1, pp. 101–17.

Rocky Mountain Institute 2007. RMI's approach to water management: creating a 'soft path' for water services. Available at www.rmi.org.

Rokosh, D.A., Chong-Kit, R., Lee, J., Mueller, M., Pender, J., Poirier, D. and Westlake, G.F. 1997. Toxicity of freeway storm water. J.S. Goudey, S.M. Swanson, M.D. Treissman and A.J.Niimi (eds) *Proceedings of the 23rd Annual Aquatic Toxicity Workshop*, 7–9 October 1996, Calgary, Alberta, pp. 151–9.

Rozanov, B.G. 1994. Constraints in managing soils for sustainable land use in drylands. D.J. Greenland and I. Szabolcs (eds) *Soil Resilience and Sustainable Land Use*. Wallingford, UK, CAB International, Chapter 10, pp. 145–53.

Ruiter, W. 1990. Watershed: flood protection and drainage in Asian Cities. *Land and Water International*, Vol. 68, pp. 17–19.

Salas, S., Heifetz, R. and Barret, E. 1990. Intestinal parasites in Central American immigrants in the USA. *Arch. Intern. Med.*, Vol. 150, pp. 1514–17.

Sansonetti, P.J. 1991. Genetic and molecular basis of epithelial cell invasion by *Shigella* species. *Rev. Infect. Dis.*, Vol. 13, pp. 285–92.

Schilling, W. 1989. *Real Time Control of Urban Drainage Systems: the state of the art.* IAWPRC Scientific Report No.2. London, Pergamon.

Schönning, C. and Stenström, T.A. 2005. Guidelines for the safe use of urine and faeces in ecological sanitation. *Proceedings of the 3rd International Ecological Sanitation Conference*, 23–26 May, Durban, South Africa.

Schueler, T.R. 1987. *Controlling Urban Runoff: a practical manual for planning and designing urban BMPs.* Washington, D.C., Metropolitan Washington Water Resources Planning Board.

Schulman, A. 1987. Non-Western patterns of biliary stones and the role of ascariasis. *Radiology*, Vol. 162, pp. 425–30.

Sharma, N.P., Damhang, T., Gilgan-Hunt, E., Grey, D., Okaru, V. and Rothberg, D. 1996. African water resources challenges and opportunities for sustainable development. Washington, D.C., World Bank. Technical Paper 331

Shilton, A. (ed.). 2005. *Pond Treatment Technology.* London, IWA Press.

Silveira, A.L.L. 2002. Problems of modern urban drainage in developing countries. *Wat. Sci. Tech.*, Vol. 46, No. 7, pp. 31–40.

Smit, J. and Nasr, J. 1992. Urban agriculture for sustainable cities: using wastes and idle land wand water bodies as resources. *Environment and Urbanization*, Vol. 4, No. 2, pp. 141–52.

Stephenson, D. 1996. Evaluation of effects of urbanization on storm runoff. F. Sieker and H.-R. Verworn (eds) *Proceedings of the 7th International Conference on Urban Storm Drainage*, Hannover, Germany, 9–13 September 1996, pp. 31–6.

Strauss, M., Drescher, S., Zurbrügg, C.H. and Montangero, A. 2003. Co-composting of faecal sludge and municipal organic waste. *A Literature and State-of-knowledge Review.* Swiss Federal Institute of Environmental Science and Technology (EAWAG), International Water Management Institute (IWMI).

Tellez, A., Linder, E., Meyer, E. and Morales, W. 1997. Prevalence of intestinal parasites in the human population of León, Nicaragua. *Acta Tropical*, Vol. 66, No. 3, pp. 119–25.

Ternes, T.A. and Joss, A. (eds) 2006. *Human Pharmaceuticals, Hormones and Fragrances: the challenge of micropollutants in urban water management*. London, IWA Press.

Thomas, J., Gibson, G., Darboe, M., Dale, A. and Waever, L. 1992. Isolation of *Helicobacter pylori* from human feces. *Lancet*, Vol. 340, pp. 1194–5.

Thompson, R.D. 1975. *The climatology of the arid world*, Geography Papers – GP#35, University of Reading, UK.

Todd, D.K. 1980. *Groundwater Hydrology*, 2nd edn. New York, John Wiley.

Todd, N.J. and Todd, J. 1993. *From Eco-cities to Living Machines: principles of ecological design*. Berkeley, Calif., North Atlantic Books.

Tokun, A. 1983. Current status of urban hydrology in Nigeria. *Proc. Urban Hydrology*, Baltimore, May/June 1983, published by ASCE New York, pp. 193–207.

Tsanis, I.K., Wu, J. and Marsalek, J. 1995. Feasibility of modeling remedial measures for micro-biological pollution of the St Clair River at Sarnia Bay. *J. Great Lakes Res.*, Vol. 21, No. 1, pp. 138–54.

Tucci, C.E.M. 1991. *Flood Control and Urban Drainage Management*. A web publication, available at: http://www.cig.ensmp.fr/~iahs/maastricht/s1/TUCCI.htm

Tucci, C.E.M. (ed.) 2001. *Urban Drainage in Specific Climates: URBAN drainage in humid tropics*, Vol. 1. Technical reports in Hydrology, No. 40, Vol. 1. Paris, UNESCO.

Tucci, C. and Villanueva, A. 2004. Land use and urban floods in developing countries. A. Szollosi-Nagy and C. Zevenbergen (eds) *Urban Flood Management*. London, Taylor and Francis.

Ujang, Z. and Henze, M. 2006. *Municipal Wastewater Management in Developing Countries: principles and engineering*. London, IWA Press.

United Nations Centre for Human Settlements (UN-HABITAT) 2003. *Water and Sanitation in the World's Cities: local action for global goals*. United Nations Human Settlements Programme. London, Earthscan/James and James.

United Nations Educational, Scientific and Cultural Organization (UNESCO) and GTZ 2006. *Capacity Building for Ecological Sanitation*. Paris, UNESCO IHP Program.

United Nations Environmental Programme (UNEP) and WHO 1992. *Urban Air Pollution in Megacities of the World*. Oxford, Blackwell.

US Department of Agriculture, Soil Conservation Service (SCS) 1975. *Urban Hydrology for Small Watersheds*. Technical release No. 55, Washington, D.C., USDA.

US Environmental Protection Agency (EPA) 1983. *Results of the Nationwide Urban Runoff Program Volume I – Final Report*. Washington, D.C., Water Planning Division, US EPA.

US Environmental Protection Agency (EPA) 1992. *Guidelines for Water Reuse*. EPA/625/R-92/004., Cincinnati, Ohio, Center for Environmental Research Information, US EPA.

US Environmental Protection Agency (EPA) 1997. *Land Application of Sludge Manual*. Columbus, Ohio, EPA.

US Environmental Protection Agency (EPA) 1999. *Biosolids Generation, Use and Disposal in the United States*. Report EPA530-R-99-009, September. Washington, D.C., US EPA.

Urbonas, B. 1994. Assessment of stormwater BMPs and their technology. *Wat. Sci. Tech.*, Vol. 29, No. 1–2, pp. 347–53.

Urbonas, B. and Benik, B. 1995. Stream stability under a changing environment. E.E. Herricks (ed.) *Stormwater Runoff and Receiving Systems*. Boca Raton, Fla., CRC Press/Lewis Publishers.

Van Blarcum, S.C., Miller, J.R. and Russell, G.L. 1995. High latitude river runoff in doubled CO_2 climate. *Climatic Change*, Vol. 30, pp. 7–26.

Van Buren, M.A., Watt, W.E. and Marsalek, J. 2000. Thermal enhancement of stormwater runoff by paved surfaces. *Wat. Res.*, Vol. 34, No. 4, pp. 1359–71.

van de Ven, F.H.M. 1988. Water balances of urban areas. In: *Urban Water 88, Proceedings of the International Symposium on Hydrological Processes and Water Management in Urban Areas*. Koblenz, Germany, IHP/OHP Secretariat, Bundesanstalt fur Gewasserkunde.

van Eyck, K., Thoeye, C., De Heyder, B. and Angelakis, A. 2001. *Management of Ongoing and Planned Water Reuse Related Projects in USA and Mediterranean Countries*. WHO Expert consultation on Health risks in Aquifer Recharge by Recycled Water. Budapest, Hungary, 8–9 November.

Viessman, W., Lewis, G.L. and Knapp J.W. 1989. *Introduction to Hydrology*, 3rd edn. New York, Harper and Row.

Viklander, M., Marsalek, J., Malmquist, P.-A. and Watt, W.E. 2003. Urban drainage and highway runoff in cold climates: conference overview. *Wat. Sci. Tech.*, Vol. 48, No. 9, pp. 1–10.

Wagner, I., Marshalek, J. and Breil, P. (eds) 2007. *Aquatic Habitats in Sustainable Urban Water Management: Science, Policy and Practice*. Leiden, Taylor and Francis/Balkema.

Walski, T.M., Chase, D.V., Savic, D., Grayman, W., Beckwith, S. and Koelle, E. 2003. *Advanced Water Distribution Modeling and Management*. Waterbury, Conn., Haested Press.

Wani, N.A. and Chrungoo, R.K. 1992. Biliary ascarisis: surgical aspects. *World J. Surg.*, Vol. 16, pp. 976–979.

Water Pollution Control Federation (WPCF) 1989. *Combined Sewer Overflow Pollution Abatement*. Manual of Practice FD-17, WPCF. Alexandria, Va., Water Environment Federation.

WaterWorld and Water and Wastewater International 2000. *World of Water: the past, present and future*. Tulsa, Okla., Pennwell Press.

Wegelin-Schuringa, M. 1999. *Water Demand Management and the Urban Poor*. Delft, the Netherlands, IRC International Water and Sanitation Centre, Special Paper (available on the Internet).

Wilderer, P.A. 2004. Applying sustainable water management concepts in rural and urban areas: some thoughts about reasons, means and needs. *Wat. Sci. Tech.*, Vol. 49, No. 7, pp. 8–16.

Wilson, L., Amy, L., Gerba, C., Gordon, H., Johnson, B. and Miller, J. 1995. Water quality changes during soil aquifer treatment of tertiary effluent. *Wat. Environ. Res.*, Vol. 67, No. 3, pp. 371–6.

Winblad, U. and Simpson-Hebert, M. (eds) 1998. *Ecological Sanitation*. Stockholm Environment Institute, Stockholm.

Wishmeier, W.H. and Smith, D.D. 1965. *Predicting Rainfall-erosion Losses from Crop-land East of the Rocky Mountains*, Agric. Handbook 282. Washington, D.C., US Dept. of Agriculture.

Wishmeier, W.H., Johnson, C.B. and Cross, B.V. 1971. A soil erodibility nomograph for farmland and construction sites. *J. Soil Wat. Cons.*, Vol. 26, pp. 189–93.

Wolf, G. and Gleick, P.H. 2003. The soft path for water. P.H. Gleick (ed.) *The World's Water 2002–2003*. Washington, D.C., Island Press, pp. 1–32.

Wolfe, P. 2000. *History of Wastewater*. World of Water 2000, Supplement to Penn Well Magazine. Tulsa, Okla., Pennwell Press, pp. 24–36.

Wolman, M.G. and Schick, A.P. 1962. Effects of construction on fluvial sediment: urban and suburban areas of Maryland. *Wat. Resour. Res.*, Vol. 3, pp. 451–64.

World Bank 1998. *World Development Report*. Washington, D.C., World Bank.

World Health Organization (WHO) 1988. *Urbanization and its Implications for Child Health: potential for action*. Geneva, WHO.

World Health Organization (WHO) 1989. *Health Guidelines for the Use of Wastewater in Agriculture and Aquaculture*. Report of a WHO Scientific Group, WHO Technical Report Series 778. Geneva, WHO.

World Health Organization (WHO) 1997. *Amoebiasis: an expert consultation*. Weekly Epidemiological Record No. 14, April. Geneva, WHO.

World Health Organization (WHO) 2003. *Guidelines for Safe Recreational Water Environments. Vol. 1. Coastal and Fresh Waters*. Geneva, WHO.

World Health Organization (WHO) 2004. *Guidelines for Drinking-Water Quality*, 3rd edn. Geneva, WHO.

World Health Organization (WHO) and UNICEF 2000. *Global Water Supply and Sanitation Assessment: 2000 report*. Geneva, WHO.

Yates, M., Gerba, C.H. and Kelly, L. 1985. Virus persistence in groundwater. *Applied and Envir. Microbiol.*, Vol. 49, No. 4, pp. 778–81.

Yu, H., Tay, J.-H. and Wilson, F. 1997. Sustainable municipal wastewater treatment process for tropical and subtropical regions in developing countries. *Wat. Sci. Tech.*, Vol. 35, No. 9, pp. 191–8.

Yusop, Z., Tan, L.W., Ujang, Z., Mohamed, M. and Nasir, K.A. 2004. Runoff quality and pollution loadings from a tropical urban catchment. CD-ROM, *Proceedings of the 8th International Conference on Diffuse/Nonpoint Pollution*, Kyoto, 24–29 October.

Zhu, K., Ning, H., Xie, C. and Zhang, L. 2006. Quality of harvested rainwater in Gansu and valid disinfection alternative. CD-ROM, *Proceedings of the IWA World Water Congress and Exhibition*, Beijing, China, 10–14 September.

Zukovs, G. and Marsalek, J. 2004. CSO treatment technologies. *Wat. Qual. Res. J. Canada*, Vol. 39, No. 4, pp. 439–48.

Index

acid rain 74–75
acidification 89, 90, 107
aesthetics 37, 47, 54
agricultural irrigation 29, 51, 61, 63–64, 96
air pollution 9, 10, 12, 67, 74–75, 90
air temperatures 10, 12, 13–14, 26, 67, 74
anthropogenic influences 3, 12, 13, 71, 78
application of biosolids 96, 98
aquatic weeds 106, 107
aquifer 3, 14, 16, 38, 60, 61, 63, 67, 69, 100–103
aquifer storage recovery systems (ASRS) 102
arid climate 11, 35, 51, 100
aridity 11
ascariasis 84

bacteria 35–36, 49, 57, 72, 81–82, 84, 85, 86, 89, 111
bacteria – indicator 44, 49, 84, 85–86
bathing 26, 77, 79, 86
benthic communities 105
benthos 104
biochemical oxygen demand (BOD) 49, 79
biodegradable organic matter 49, 101
biodiversity 78, 103–107, 111
biofilters 46, 47
biofiltration 45, 46
bioretention 45, 47
biosolids 43, 56, 94, 96, 98
biota 7, 38, 75, 88, 89, 103–107, 111
blackwater 56, 59
BMPs (best management practices) 45, 46, 47
bottled water 28, 30, 37
brine 35
buffers 46, 89

canalization 70, 71
catchment 4, 6, 12, 13, 16–17, 21, 39–40, 47, 54, 69–70, 76, 94
centralized wastewater systems 52
chemical effects 79–81
chemical oxygen demand (COD) 79
chemical substances 6, 44, 46, 48–49, 56, 58, 80, 93, 97, 99, 106
chloride 33, 44, 77, 78, 80, 90, 93
chlorination 24, 35, 50, 64
chlorine 35, 63, 80, 83
chlorine residual 35
climate 10–11, 48, 60, 67, 77, 89, 90, 91
climate change 2, 4, 12, 41, 78
coastal waters 24, 73, 75
cold climate drainage 47, 48, 58
combined sewer system 48, 50, 94
combined sewer overflows (CSOs) 38, 43, 48–50, 68, 73, 79, 80, 84, 85, 86
CSO abatement 50
CSO characterization 48–49
CSO control 49–50
CSO storage 49
CSO storage – off-line 49–50
CSO treatment 49–50
cumulative effects 72

demand management 7, 25, 26–27, 110
density stratification 77–78, 106
depression storage 5, 13
desalination 28, 33–35, 61
diarrheic disease 53
diseases 20, 24, 35, 38, 53, 54, 62, 81, 86
disease vectors 93, 98
disinfection 24, 33, 35–36, 54, 81, 110
disinfection – solar 35
dissolved organic carbon (DOC) 88
dissolved oxygen (DO) 44, 72, 79, 80, 88–89, 104

Milton Keynes UK
Ingram Content Group UK Ltd.
UKHW050451071024
449327UK00015B/324